D1038988

SOUTHEASTERN UNITED STATES

showing locations of generalized physiographic provinces

The Ridge and Valley, Blue Ridge, and
Piedmont jointly form the Appalachians.

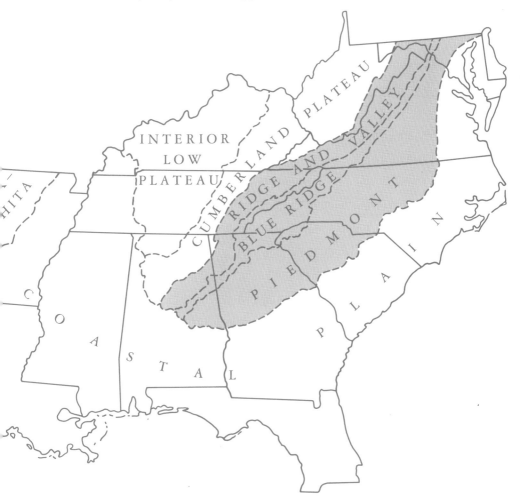

Trees of the
Southeastern United States

WORMSLOE FOUNDATION PUBLICATIONS

NUMBER EIGHTEEN

Trees of the Southeastern United States

Wilbur H. Duncan
and Marion B. Duncan

The University of Georgia Press
Athens and London

The University of Georgia Press paperback edition, 2000
© 1998 by The University of Georgia Press
Athens, Georgia 30602
All rights reserved

Set in Linotron 202 10 on 11 Garamond Number Three

The paper in this book meets the guidelines for permanence
and durability of the Committee on Production Guidelines
for Book Longevity of the Council on Library Resources.

Printed in China

11 10 09 08 07 P 7 6 5 4 3

The Library of Congress has cataloged the hardcover edition
of this book as follows:

Library of Congress Cataloging-in-Publication Data
Duncan, Wilbur Howard, 1910–
Trees of the southeastern United States / Wilbur H. Duncan
and Marion B. Duncan.
xi, 322 p. : ill. (some col.), maps ; 24 cm. —
(Wormsloe Foundation publications ; no. 18)
Maps on lining papers. Includes index.
ISBN 0-8203-0954-0 (alk. paper)
1. Trees — Southern States — Identification. I. Duncan, Marion B.
II. Title. III. Series: Publications (Wormsloe Foundation) ; no. 18.
QK125 .D79 1988   582.160975—dc19   87-5837

Paperback ISBN-13: 978-0-8203-2271-1 — ISBN-10: 0-8203-2271-7

British Library Cataloging-in-Publication Data available

For Mack, Lucia, and Douglas

Contents

Acknowledgments

This volume represents the fruits of countless pleasant and productive field trips over several decades with students, colleagues, friends, and family. During this time we have enjoyed continuous support from the Department of Botany, University of Georgia. We wish to give special recognition to Dr. James Hardin, North Carolina State University, who has encouraged this effort with personal loyalty as well as professional aid. We are grateful to David Hunt for his help with the genus *Quercus* and to the following friends who have generously granted us use of their photographs: Michael Dirr (*Elaeagnus angustifolia, Cotinus obovatus, Leitneria floridana, Malus ioensis, Melia azedarach*); Fred Galle (*Viburnum obovatum*); John Garst (*Magnolia macrophylla*); Jim Hamrick (*Carya texana*); Ron Lance (*Carya laciniosa, C. myristiciformis, Euonymus atropurpureus, Ilex ambigua, I. coriacea, I. montana, Pinus serotina, Populus heterophylla, Prunus avium, Sophora affinis*); D. A. Rayner (*Ilex amelanchier*); Carol Ruckdeschel (*Manihot grahamii*); W. M. Shephard (*Quercus breviloba*); and William Mahler (*Prunus mexicana*).

Introduction

This manual is a guide for the identification of trees reproducing naturally in the southeastern United States. It also provides pertinent information about each. Color photographs that focus on recognition characteristics are provided for 276 species; the photographs are accompanied by descriptive texts for each species. Occasional secondary paragraphs describing similar species bring to 306 the total number of species treated. Keys are included as aids in identification, based on the most useful and easily recognized diagnostic vegetative, flowering, or fruiting characters, or combinations of these.

The Southeast is defined to include Alabama, Mississippi, Louisiana, Arkansas, Tennessee, Kentucky, West Virginia, Maryland, Delaware, Virginia, North Carolina, South Carolina, Georgia, and that portion of Florida from Levy County on the Gulf Coast to Flagler County on the Atlantic and northward. Peninsular Florida is omitted because it is generally a freeze-free area supporting a large assemblage of subtropical species native or adapted to the area and strikingly different from the flora of the other states of the Southeast. These artificial borders do not preclude the appearance of some species beyond them; indeed, because of many range extensions, this manual can be useful in adjacent states and often beyond. Mainly for this reason, the complete range for each species is indicated. A map inside the front and back covers indicates the area involved, other eastern states and provinces, and a generalized representation of the physical regions of the Southeast. One or more of these regions may be referred to in a species description, emphasizing the area of its occurrence.

A distribution map accompanies most of the descriptions and is based on our observations, communications with specialists, specimens in herbaria, and reliable published data. Distributions extending beyond the area covered by the map are given in the texts.

How to Use This Manual

What Is a Tree?

A tree, as the term is used here, is a plant with a perennial stem of at least 4 meters (13 feet) in height and a minimum diameter of 7.5 centimeters (3 inches) at breast height (DBH), measured at 1.37 meters (4.5 feet) above the ground. The definition derives from an artificial distinction between trees and shrubs that is based on these two dimensions. Any species known to us as having one or more individuals that have reproduced naturally in the Southeast and have reached tree size is included. If a species produces shrubby individuals that might be encountered in the Southeast, this tendency is indicated in the description of that species.

How to Use the Keys

This manual separates trees of the southeastern United States into eleven major categories (groups) based on broadly distinctive characteristics, such as narrowly linear leaves rather than broad ones, simple leaves versus compound, or evergreen leaves as opposed to deciduous ones. Each of these eleven categories, designated Group A through Group K, brings together kinds of trees that have certain characteristics in common and provides a dichotomous key for determining the genus to which a specimen belongs. Where a family or genus has members representative of more than one group, the family or genus is divided and treated in all appropriate groups. For example, the legume family, Fabaceae, has genera with compound leaves placed in Group D and a genus with simple leaves in Group G. The genus *Ilex* has species in Group H with the evergreen plants and also in Group K, one of the groups with deciduous leaves.

The first step in an orderly approach to identification is, therefore, to determine the group into which the specimen in question falls. Help in making this choice is provided by a dichotomous key to the groups on page 22. Those unfamiliar with the use of dichotomous keys may wish to refer to an explanation of this use on page 3.

The second step is to decide to what genus in that group the specimen belongs. A key to genera is provided for this determination. If only one species in the genus chosen is involved, then the search is ended. If two or more species are involved, a third step is required in choosing from among those species, and a key is provided for this purpose.

The choice of a species made by the above three steps is actually only a tentative one. Both the photograph and the accompanying description should be checked carefully against the specimen. Checking the distribution may sometimes prove helpful. If unsure of a term used, one should check the glossary and the illustrations of tree characteristics provided elsewhere. One should also keep in mind that a species may vary considerably, a single photograph does not include all such variations, and a particular specimen may be one of those variants not shown in the photograph. It is a good practice, when possible, to check several parts of the same tree, because the specimen chosen could be unusual. The descriptions take such variations into account, with rare exceptions, and should allow one to determine whether the correct species has been reached. When in doubt, one should return to the key to groups, checking each step more carefully. If confident of the original choice of a group—and making choices should become easier with time—then one could return directly to the key to genera of that group and recheck for the genus. If confident of the original choice of a genus, one may then simply recheck for the species.

Dichotomous Keys

A dichotomous key is arranged in couplets that offer a choice between two opposing characters or sets of characters, each of the two headed by the same number. For example: "1. Leaves rounded at the tip," and somewhere below, "1. Leaves pointed at the tip." By making a choice of one of these statements, the user either finds the proper group, genus, or species, or is directed to another couplet for another choice. This is repeated until the correct group, genus, or species is determined. Should a wrong choice be made in a set of characters, there is no way to reach the correct conclusion without starting over at least part way, because there are no crossroads for getting back on the proper track. Such an error becomes apparent when one reaches a couplet in which the characters do not apply or the description does not match. The advantage of a dichotomous key is that it focuses on only one or a few easily accessible characters at a time and avoids the time-consuming check of all the characters in a group, genus, or species.

Species Names and Descriptions

Each species is described under its family heading and is introduced first by its common name or names, then the scientific name, which is in Latin, and the authority or authorities who first described and named it. Common names often vary widely from one locality to another and thus are unreliable for accurate usage. They are included here as a possible means of familiar reference and were chosen as the ones that appeared to be most frequently used within the geographic range involved. Scientific names of species are always binomials, that is, they are composed of two words, the genus and the specific epithet. They are always in Latin, they must comply with the International Code of Botanical Nomenclature, and no two kinds of plants may bear the same name. Because this system is recognized internationally, scientific communities throughout the world can communicate in a common terminology regardless of national language.

Taxonomic research continually reveals plant relationships previously unknown or unrecognized, sometimes with resulting synonyms. Two or more accepted species, subspecies, or varieties may be combined into one, in which case the older or oldest name is the correct one and the other name or names become synonyms. Occasionally a species is named twice; again the older name prevails. We have made a serious effort to choose valid scientific names based on current literature. Because these names sometimes vary from those used in manuals, synonyms have been included at the conclusion of some species descriptions.

Immediately under the plant name, each description begins with a listing of characters that are diagnostic for that species and make it unique. Perhaps its twigs are hairy whereas hairless twigs characterize all remaining members of that genus. In some species, only a few differentiating features may be needed; others may require a considerable number.

Following the diagnostic features unique to that species, additional descriptive characters are supplied: tree size (maximum height and DBH), vegetative features such as leaf shape or twig color, flower and fruit descriptions (pollen- and seed-bearing cones in conifers), abundance, habitat, range, elevation, and flowering period (time of pollen shedding in conifers).

Abundance is indicated as common, occasional, or rare without benefit of actual population studies by the authors. Estimates are based on current literature, field experience, herbarium records, and observations of other specialists. Common and rare are the significant ratings, occasional being not clearly different from either except that any species listed as occasional is certainly neither quite common nor quite rare. Species that appear to be abundant locally but have a restricted overall distribution are classed as rare, as are those of broadly scattered occurrence but sparse in all localities.

Those plants that we know to have any part poisonous to man or other animals through contact or ingestion are so indicated. Plant uses are touched upon lightly, with the knowledge that others may exist but also that such practices can result in species depletion or eradication. Much has been written about edible wild plants, those with medicinal qualities, species suitable for ornamental plantings, and those useful in minor capacities, as in basketry or for dye stuffs. Fortunately for those interested in obtaining plants for any of these purposes, many nurseries are now propagating trees from seeds or cuttings.

How to Use the Distribution Maps

The general distribution of a species can be determined by the size and position of dots that are placed in each state or province in which it occurs. These dots are of two sizes, the larger dot indicating that the species is relatively common in the area involved, the smaller dot that it is relatively uncommon, either because it is rare, scattered, or only locally abundant.

Dots are positioned as near as possible to the center of distribution for each state or province or portion(s) of it. If a species occurs throughout all of a state or province, or nearly so, a large dot is placed equally distant from the state boundaries, that is, in the center. If it is present in only one part, the dot is centered in the area of occurrence, as seen in the distribution map for *Pinus serotina*.

Representation of distribution sometimes requires more than one dot, such as a large dot centered where the species is relatively common and one or more small dots where uncommon, as for *Acer leucoderme* in Alabama, Georgia, and North Carolina. Two, or rarely more, large dots are used where areas of common occurrence are significantly separated, as for *Acer floridanum* in South Carolina, or where a single area not occupying an entire state or province is shaped in a manner that cannot be represented by a single dot, as in Ontario frequently and for *Ilex opaca* in Arkansas.

Since dots are centered, distinctly separated areas of distribution can be ascertained by noting distances of the dots from the margins and from each other. A single dot placed on the line separating Connecticut and Rhode Island signifies the species occurs in both states.

Photographs

Photographs accompanying the descriptions have been focused for clarity and to include as many diagnostic characters as is practical. If hairy twigs are significant for identification, displaying this feature may take precedence over a showy inflorescence or artistic composition. If the fruit is required for identification, it is usually emphasized or even shown alone. The designation × plus a number indicates magnification of the photograph. If the picture is twice actual size it is labeled × 2, actual size is × 1, and × ½ indicates a picture half the size of the actual subject.

Photographs and descriptions are designed to complement each other, and neither should be used to the exclusion of the other. *Ilex vomitoria* is pictured with red fruits, but they may be yellow, though rarely, and this must be learned from the text.

A voucher specimen was taken of most plants that were photographed. A specimen was always taken when the identity of a plant was in doubt, and the specimen was studied where adequate books and other published information and comparative specimens were available. These voucher specimens are deposited in the Herbarium, Department of Botany, University of Georgia, Athens, Georgia.

Abbreviations

A uniform system of abbreviations is utilized throughout the descriptive texts in the interest of brevity but without sacrificing clarity. All states and Canadian provinces are assigned abbreviated forms which we believe are more easily recognized by the public than those now used by the United States Postal Service. The same general format is followed in shortening the names of continents, regions, and compass directions. These all seem to be easily interpreted, but the reader may refer to the List of Abbreviations (page 15) in case of doubt.

Metric System

It is necessary to use the metric system in order to demonstrate comparative sizes of small parts. Only three measurements are used: meter (m), centimeter (cm), and millimeter (mm). A meter is 100 centimeters or 1000 millimeters. A metric scale in centimeters and millimeters is included for convenience inside the front and back covers. Conversion of meters to feet may be achieved by multiplying the metric value by 3 and to this product adding $\frac{1}{10}$ of it. Thus, 220 meters equals 660 feet plus 66, or approximately 726 feet.

Structure of Trees

Identification of trees requires an ability to perceive differences in structure between one species and another. There are general vegetative characteristics that are common to all trees, but fundamental reproductive processes and a few vegetative differences produce a major division into two categories: Conifers and Flowering Trees. The most obvious difference between the two is that conifers have no flowers or fruits; pollen and ovules (and seeds developing from them) occur in separate specialized structures, usually cones, instead of in flowers. In almost all species of conifers, ovules and seeds are attached to the surface of scales of the female cone instead of inside a pod or other fruit as in flowering plants. Uncommonly, seeds of conifers are surrounded by or are inside a fleshy cuplike structure, an aril, as in the yews. The leaves of almost all conifers are evergreen, exceptions being Bald-cypress and Larch; they rarely have stalks; and they are all simple and narrow, often being needlelike, awllike, or scalelike. Leaves of flowering plants, by contrast, are more commonly deciduous, most frequently petioled, sometimes have more than one blade, and are not needlelike, awllike, or scalelike. Tamarisk, a flowering plant with needlelike leaves, is an exception among southeastern trees.

The following diagrams illustrate various vegetative structures and the names applied to most shapes of tree leaves and to stem forms. It is useful to refer to these diagrams and also to the glossary.

Vegetative Structures

Conifers and flowering trees are alike in that the basic vegetative parts consist of root, stem, and leaf. They both bear *vegetative buds,* a structure in which small to minute leaves occur on a diminutive stem, later to develop into the stem and leaves that appear conspicuously on trees during part or all of the year. Such buds occur on most stem tips (*terminal buds*) or on the stem at the base of leaves (*axillary buds*). *Scales,* which are small leaves, cover most buds.

In flowering plants there are buds that develop into flowers only, as in Peach, and are called *flower buds*. There are also buds that develop into stems bearing both leaves and flowers, as in wild Black Cherry; these are called *mixed buds*.

Stems are usually above ground and bear leaves that drop off during the year they appear, or at the most remain alive on the stem for only a few years. When a leaf drops off, a scar is left on the stem. Close inspection of a *leaf scar* discloses *bundle scars* that are from the vein system (vascular bundle traces) running from the stem into the leaf. Each circular section of the stem that bears a leaf or leaves is a *node* and that part of the stem between two adjacent nodes is an *internode*. The main central stem of a tree is the *trunk,* which is covered with *bark*. Some species may have two or more trunks of approximately the same size, as in River Birch.

Vegetative Structures

SHAPE OF BLADES (LEAF, PETAL, SEPAL, BRACT)

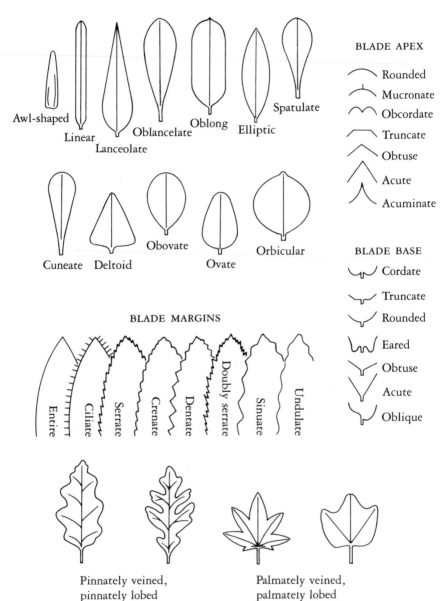

BLADE APEX

Rounded
Mucronate
Obcordate
Truncate
Obtuse
Acute
Acuminate

BLADE BASE

Cordate
Truncate
Rounded
Eared
Obtuse
Acute
Oblique

Awl-shaped
Linear
Lanceolate
Oblanceolate
Oblong
Elliptic
Spatulate

Cuneate Deltoid
Obovate
Ovate
Orbicular

BLADE MARGINS

Entire
Ciliate
Serrate
Crenate
Dentate
Doubly serrate
Sinuate
Undulate

Pinnately veined,
pinnately lobed

Palmately veined,
palmately lobed

Vegetative Structures

SIMPLE LEAVES

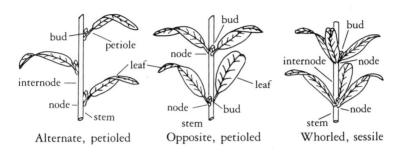

Alternate, petioled Opposite, petioled Whorled, sessile

ONCE COMPOUND LEAVES

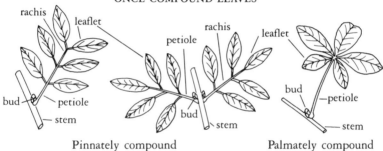

Pinnately compound Palmately compound

TWICE COMPOUND LEAF   LEAF WITH STIPULE

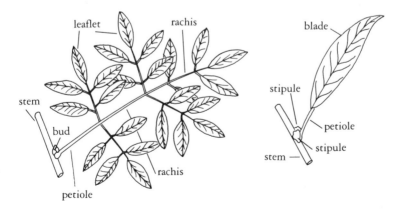

Vegetative Structures

DORMANT TWIGS

PITH FEATURES

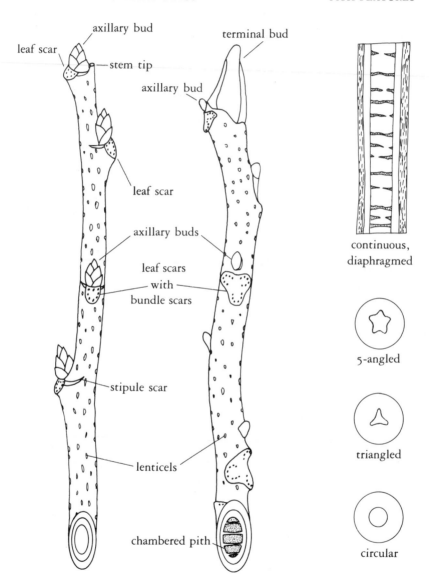

continuous,
diaphragmed

5-angled

triangled

circular

Roots are usually underground, bear no leaves, and thus have no leaf scars, nodes, or internodes. Stems sometimes sprout from roots, an important means of reproduction for some trees such as Black Locust and some of the plums.

In flowering trees, as opposed to conifers, leaves have an expanded part (the *blade* or blades), usually have a stalk (the *petiole*), and often have two *stipules*. The stipules may be fastened to the sides of the petiole, to both petiole and stem, or only to the stem near the petiole. There are always two or none, sometimes fused and appearing as one, as in Sycamore and Magnolia. They can vary in size from minute to easily seen and occasionally are modified into spines such as those on Black Locust.

Blades of a leaf may be one to many. If there is one blade, the leaf is *simple;* if there are two or more blades (*leaflets*), the leaf is *compound.* A compound leaf with sets of opposite or alternate leaflets is superficially similar to a stem with opposite or alternate leaves of one blade each. In trees, excepting conifers, the most useful way to determine "what is a leaf" is to ascertain the relative positions of leaves and axillary buds, the latter being on the stem just above the leaf base.

Flowering Structures

In this manual floral characters are given a status secondary to vegetative ones where possible, because the flowering period of many trees is brief and flowers are therefore often unavailable. If available, however, they can be most useful in the identification process, and knowledge of their structure is helpful.

A complete flower consists of four basic parts: calyx, corolla, stamens, and pistil(s). The last two are necessary for reproduction. *Calyx,* the outermost part of the flower, is composed of *sepals.* These are usually green and often inconspicuous; they may occur as separate distinct parts or they may be partially or completely fused. The *corolla,* which lies above the calyx, consists of *petals.* These are the organs that are usually colorful and attract attention; like sepals, they may be discrete, partially united, or totally fused. The corolla may be absent from a flower or, as in willows, both calyx and corolla may be missing.

Stamens are pollen-bearing structures, sometimes referred to as the male part of the flower, and may be separate or united. The *pistil,* also called female part of the flower, contains *ovules* that later become seeds. The pistil itself, and sometimes adhering structures, develops into a *fruit* that contains the seeds. Pistils may occur singly, or there may be two to many; each individual pistil may have one to several stigmas and styles, or the styles may be missing. Some tree species produce flowers with pistils sterile or lacking, or with stamens sterile or lacking. These staminate and pistillate flowers may occur on the same plant, as in alders and hickories, or on separate plants, as is true with willows and many hollies.

Calyx, corolla, stamens, and pistils are usually attached to the *receptacle,* the upper, often enlarged end of the flower stalk (the *pedicel*). In some flowers stamens are attached to the corolla; in some, sepals, petals, and stamens are supported by a *hypanthium,* a cup-, saucer-, or disc-shaped body lying between these parts and the receptacle. Positions of the flower parts relative to each other are diagrammed below; individual terms are defined in the glossary.

Floral Structures

COMPLETE FLOWERS WITH SUPERIOR OVARY AND NO HYPANTHIUM

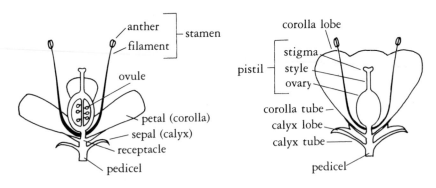

COMPLETE FLOWERS WITH HYPANTHIUM

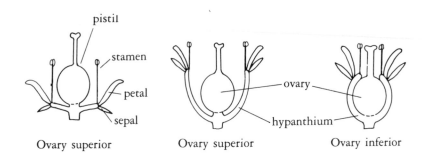

Ovary superior   Ovary superior   Ovary inferior

FLOWER FORM

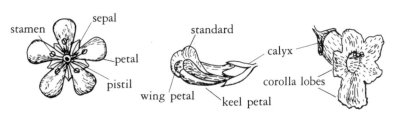

Regular   Legume, irregular   Catalpa, irregular

Inflorences (flower clusters)

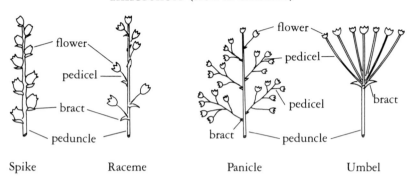

Spike   Raceme   Panicle   Umbel

Fruiting Structures

The fruit of a flowering plant develops from a pistil consisting of one or more carpels together with any other structure that may adhere to the matured pistil. Examples of fruits developing from a single carpel are plums and the seedpods of Redbud. Fruits developing from two or more carpels include those of Maple, Sourwood, and Rhododendron. Some fruits that develop from two or more carpels, as revealed by features such as the number of stigmas, have all but one carpel abort as the fruit is formed. Examples are the fruits of Oak and Ash. And then there are fruits that develop from more than one flower, as those of Mulberry, Sweetgum, and Sycamore.

The most common structure adhering to the pistil or pistils is the *hypanthium*. In apples this structure forms the outer fleshy part that is eaten. Other examples are the fruits of Dogwood, Sparkleberry, and Viburnum. In each of these four fruits sepals are or were attached at the end opposite the fruit stalk. Sometimes the sepals drop before the fruit matures, in which case scars are present where they were attached, as in Dogwood and Viburnum.

Among the various kinds of fruits, two of the most difficult to distinguish are *berry* and a *drupe* that has two or more stones. Both types are fleshy and contain one to several seeds. The seeds of both consist of an embryo, sometimes endosperm, and the seed coat, which may be thin or thick and hard, as in Persimmon. In berries each seed is immediately surrounded by the fleshy part of the fruit as in Sparkleberry and Persimmon. In drupes each seed is immediately surrounded by a hard often bony wall, the whole referred to as a pit or stone, and one or more of these is embedded in the fleshy part of the fruit. Examples of drupes with one stone are Peach and Cherry, and of those with several stones, Holly. Fortunately, for the purposes of identification it is rarely necessary to make a distinction between a berry and a drupe. Other types of tree fruits include capsules, follicles, nuts, nutlets, pomes, samaras, and more, all described in the glossary.

Note to the Amateur

It is possible to leaf through this volume in search of a photograph that matches a particular specimen, compare the specimen with the printed description, and thus arrive at a correct identification; but the method is usually slow and difficult. For those who do not yet possess a fairly thorough knowledge of trees and are not familiar with the arrangement of this book and the aids provided, a prescribed and orderly approach is likely to save much time and improve accuracy.

To provide enough information for accurate identifications of all but a few species of trees in the Southeast, it occasionally has been necessary to use technical terms and minute characters, although these have been avoided as much as practical. By relying on the photographs and the less technical parts of the descriptions, however, the amateur will be able to identify specimens just as well as if the more technical terms and minute characters had been omitted. In addition, amateurs who use this book regularly become more familiar with the terminology and structures, and will begin to find the technical information more helpful.

Abbreviations

Ala	Alabama	NWT	Northwest Territories
Alas	Alaska	Okla	Oklahoma
Alta	Alberta	Ont	Ontario
Ariz	Arizona	PEI	Prince Edward Island
Ark	Arkansas	S. Amer	South America
C. Amer	Central America	Sask	Saskatchewan
Cal	California	SD	South Dakota
Can	Canada	SE	southeastern region
Co.	county	Tenn	Tennessee
Col	Colorado	Tex	Texas
Conn	Connecticut	Va	Virginia
CP	Coastal Plain	Wash	Washington
Fla	Florida	W. Indies	West Indies
Ga	Georgia	Wisc	Wisconsin
Ill	Illinois	WVa	West Virginia
Ind	Indiana	Wyo	Wyoming
Ky	Kentucky		
La	Louisiana	c	central
Lab	Labrador	e	eastern
Man	Manitoba	n	northern
Mass	Massachusetts	ne	northeastern
Md	Maryland	nw	northwestern
Me	Maine	s	southern
Mex	Mexico	se	southeastern
Mich	Michigan	sw	southwestern
Minn	Minnesota	w	western
Miss	Mississippi		
Mo	Missouri	cm	centimeter
N. Amer	North America	m	meter
NB	New Brunswick	mm	millimeter
NC	North Carolina		
ND	North Dakota	f.	Symbol for a taxonomic form.
NE	northeastern region		Also indicates son of an author.
Neb	Nebraska	sp.	Symbol for a single species.
Nev	Nevada	spp.	Symbol for more than one
Nfld	Newfoundland		species.
NMex	New Mexico	ssp.	Symbol for a subspecies.
NS	Nova Scotia	var.	Symbol for a variety.

Glossary

ACUMINATE.   Tapering gradually to a long thin point. Compare with ACUTE.

ACUTE.   Applied to tips and bases of structures ending in a point less than right angle. Compare with ACUMINATE and OBTUSE.

ALTERNATE.   One leaf, bud, or branch per node.

APEX.   Tip; end opposite point of attachment.

APICULATE.   Ending with a short sharp abrupt point.

APPRESSED.   Lying flat against.

ARIL.   Fleshy or pulpy covering or appendage from the base of a seed.

ARILLATE.   Having an aril or arils.

AURICULATE.   Eared.

AWL-SHAPED.   Having a linear shape and tapering to a fine point; narrowly triangular.

AXIL.   The space between any two adjoining organs, such as stem and leaf.

AXILLARY.   In an axil.

BERRY.   Any fruit with fleshy walls and with a few to many seeds encased in soft tissue, such as Grape, Tomato, and Blueberry. See DRUPE.

BISEXUAL.   Having both sexes present and functional in the same individual.

BLADE.   The flattened and expanded part of a leaf, or parts of a compound leaf.

BRACKISH.   Somewhat salty.

BRACT.   A reduced leaf, particularly at base of flower or flower stalk.

BRANCHLET.   A small or secondary branch.

BRISTLE.   A stiff hairlike structure.

BUNDLE SCAR.   Scar within a leaf scar where the vein system broke when the leaf dropped.

BUR.   A rough, prickly, or spiny flower or fruit (or cluster of either) and any associated parts.

CALLUS, CALLOSITY.   A localized hardened or thickened protuberance or prominence.

CALYX.   Collective term for all the sepals of a flower, whether separate or united; the outer series of flower parts and of the perianth.

CAPSULE.   A dry fruit with two or more rows of seeds, i.e., with two or more carpels, the fruit splitting open (dehiscing) at maturity.

CARPEL.   A simple pistil or a unit of a compound pistil.

CATKIN.   Inflorescence consisting of numerous bracts and unisexual apetalous flowers densely arranged on drooping or erect spikes, as in willows and oaks.

CILIATE.   Fringed with hairs. Can be said of any hairy margin, whether entire, toothed, lobed, or otherwise.

CLAW.   The narrowed parallel-sided base of a sepal or petal in some kinds of flowers.

COASTAL PLAIN.   Dry-land portion of the Atlantic Plain as opposed to the Continental Shelf, the underwater portion. It slopes gradually seaward and includes barrier islands.

COMPOUND LEAF.   One divided into 2 or more blades (leaflets). *Palmately compound* leaves have 3 or more leaflets arising from a common point. *Pinnately compound* leaves have leaflets arranged along a common axis.

CONE.   A mass of ovule-bearing or pollen-bearing bracts or scales that are arranged spirally on a cylindrical or globose axis; common to most conifers.

CONIC.   Cone-shaped.

CONIFER.   Plants without flowers, the ovules naked. Any of an order of trees and shrubs bearing true cones (pines) or with arrillate seeds (yews).

CORDATE.   Heart-shaped in outline; descriptive of an ovate organ (such as a leaf) with two rounded basal lobes.

CORIACEOUS.   Having a stiff leathery texture.

COROLLA.   Collective term for all the petals of a flower, whether separate or united; the inner series of the perianth. If there is only one series, it is the calyx and the corolla is absent.

CORYMB.   A flat- or round-topped flower cluster in which the outer pedicels are longer than the inner ones, the outer flowers opening before the center ones.

CRENATE.   Having margins scalloped with shallow rounded teeth.

DBH.   Diameter breast high. Generally accepted standard for measuring trees at 4.5 feet or 137 cm above ground.

DECIDUOUS.   Not evergreen; foliage dying and usually falling after growing season.

DECURRENT.   Extending down and attached to the stem, forming a ridge or wing.

DEHISCENT.   Opening by natural splitting, as an anther discharging pollen or a fruit its seeds.

DENTATE.   Toothed; having sharp spreading coarse teeth that are perpendicular to the margin. Compare with SERRATE.

DENTICULATE.   Finely dentate.

DIAPHRAGM.   A cross partition; a membrane that separates, as in pith.

DRUPE.   Fleshy indehiscent fruit, such as Cherry or Plum, having a single seed encased in a hard stony covering, or sometimes with more than one such encased seed, as in Holly.

DUNE.   Pile or ridge of loose sand deposited by wind action.

EARED.   Having an earlike process or appendage, usually at the base of an organ; frequently applied to leaves and petals. AURICULATE.

ELLIPSOID.   Said of a 3-dimensional body whose plane sections are all either ellipses or circles.

ELLIPTIC.   Oblong with the ends equally rounded or nearly so.

ENTIRE.   Smooth, without teeth or indentations; applied to margins, edges.

EVERGREEN.   Plants with live leaves persisting through one or more winter seasons.

FALCATE.   Crescent-shaped.

FIMBRIATE.   Margin divided into narrow or filiform segments often of irregular sizes. Fringed.

FOLLICLE.   A dry one-celled fruit with a single placenta and splitting along the opposite edge.

FRUIT.   A matured pistil together with any other structure that may adhere to it; the seed-bearing organ of a flowering plant and any attached structure.

GLABROUS.   Without hairs, bristles, or stalked glands.

GLAND.   A depression, protuberance, or appendage on the surface of an organ, which secretes a usually sticky fluid; any structure resembling such a gland.

GLAUCOUS.   A surface with a fine white substance (bloom) that will rub off, as on some grapes and blueberries.

GLOBOSE.   Globe-shaped; spherical.

GLUTINOUS.   Sticky, gummy, having quality of glue.

HAMMOCK.   A raised fertile area in the midst of a wetland and characterized by hardwood vegetation and deep humus soil.

INDEHISCENT.   Not opening naturally at or after maturity.

INFLORESCENCE.   Any complete flower cluster including branches and bracts. Clusters separated by leaves are separate inflorescences.

KEEL.   A central ridge on the back of plant parts such as sepals, petals, or bud scales. The two lower adhering petals of some legumes.

KNEE.   A woody knob arising from the root and extending above ground or water level.

LANCEOLATE.   Much longer than wide, widest below the middle, tapering toward apex, or both apex and base; resembling a lance head.

LEAF SCAR.   Mark left on a twig where a leaf has broken off.

LEAFLET.   A single segment (blade) of a compound leaf.

LENTICEL.   Small corky spot or line on the bark of a twig, branch, and/or trunk of a woody plant.

LINEAR.   Narrow and elongated with sides parallel or nearly so.

LOBE.   Segment of a leaf between indentations that do not extend to the midrib or base of the leaf.

MIDRIB.   Central or main vein of a leaf or leaflike part.

NEEDLE.   A narrow usually stiff leaf, as in pines, firs, and hemlocks.

NODE.   The narrow region on a stem where a leaf or leaves are or were attached.

NUT.   An indehiscent one-seeded fruit having a hard outer wall.

NUTLET.   A small nut loosely distinguished by its size.

OB-.   A prefix signifying an inversion, such as obcordate, the opposite of cordate.

OBLONG.   Elongate in form with sides parallel or nearly so, the ends more or less blunted and not tapering; wider than LINEAR.

OBTUSE.   A blunt point, the angle of the point being greater than 90 degrees. Compare with ACUTE.

OPPOSITE.   Growing in pairs, one on each side of the axis and 180 degrees from each other.

ORBICULAR.   Circular in outline.

OVARY.   The part of the pistil containing the ovules, which develop into seeds; matures into a fruit. Also called OVULARY.

OVATE.   A 2-dimensional structure having the outline of an egg with the wider half below the middle.

OVOID.   A 3-dimensional structure having the shape of an egg with the broader half below the middle.

OVULE.   The egg-containing structure that, after fertilization, develops into a seed. In flowering plants, in the ovary; in conifers, naked on the surface of cone scales or otherwise exposed.

PALMATE.   Radiately arranged, ribbed, or lobed, as fingers of a hand.

PANICLE.   An irregularly compound raceme.

PAPILLOSE.   Bearing small nipplelike projections.

PEDICEL.   The stalk of a single flower.

PEDUNCLE.   The main flower stalk of the inflorescence supporting either a cluster of flowers or the only flower of a single-flowered inflorescence.

PELTATE.   Having the stalk of a leaf attached to the lower surface of the blade somewhere within the margin rather than on the margin. Also descriptive of stalked scales.

PENDULOUS.   Drooping or hanging loosely.

PERFECT.   Flowers having both functional stamens and pistils.

PERIANTH.   The calyx and corolla collectively, or the calyx alone if the corolla is absent.

PERSISTENT.   Remaining attached past expected time for dropping.

PETAL.   One of the parts of the corolla, the inner set of the perianth; may be separate or united to another petal.

PETIOLE.   The attaching stalk of a leaf; sometimes absent.

PINNATE.   Having lobes or blades of a leaf arranged along the sides of a common axis, as the pinnae of a feather. Also applies to major lateral veins of a leaf.

PISTIL.   The female ovule-bearing organ of a flower, composed of stigma(s) and ovary, usually with a style or styles between; consists of a single carpel or of two or more fused carpels.

PIT.   Sometimes used for STONE.

PITH.   Soft spongelike tissue at center of woody stems. *Chambered pith* consists of air spaces separated by cross partitions (diaphragms). *Continuous pith* is uninterrupted by air pockets or chambers. *Diaphragmed pith* is a continuous pith with firm cross partitions.

PLACENTA.   The ovule-bearing surface in the ovary and seed-bearing surface in the fruit.

POD.   Any dry dehiscent fruit.

POLLEN.   The male sporelike structures produced by anthers in flowers and by male cones of pines and their relatives.

POME.   A fleshy fruit, as in Apple or Pear, having several seed chambers formed from the ovary wall, part of which is parchmentlike or bony in texture; the fleshy portion formed largely from the hypanthium.

PRICKLE.   A small sharp spinelike projection that is part of the bark or epidermis as in Hercules-club. See THORN, SPINE.

PUNCTATE.   Spotted with colored or translucent dots or depressions, usually due to glands.

RACEME.   An inflorescence in which stalked flowers are arranged singly along a common elongated axis.

RACHIS.   The main axis of a spike; or of a pinnately compound leaf, excluding the petiole.

REFLEXED.   Abruptly turned or bent toward the base. See RETRORSE.

RESIN.   Sticky plant exudate insoluble in water; especially in pine and fir.

RETRORSE.   Directed backward and downward. See REFLEXED.

REVOLUTE.   Rolled under at the margin, i.e., toward the underside.

ROSIN.   Hard substance remaining after evaporating turpentine from pine resin.

SAMARA.   Winged indehiscent fruit with one seed, as in elms; paired in maples.

SCABROUS.   Rough or harsh to the touch due to minute stiff hairs or other projections.

SCALE.   Applied to many kinds of small thin flat appressed usually dry leaves or bracts, often vestigial. Sometimes epidermal outgrowths, if disclike or flattened.

SCURFY.   Surface with small scalelike or branlike particles.

SEED.   A ripened ovule containing an embryo capable of producing a new plant.

SEPAL.   One of the parts of a calyx or outer set of flower parts; may be separate or united to another sepal.

SERRATE.   Having sharp sawlike teeth pointed upward or forward.

SESSILE.   Without any kind of stalk.

SHEATH.   A tubular structure surrounding an organ or part, such as the basal part of a leaf; the circle of scales around the base of pine needles.

SHRUB.   A woody plant under tree size, frequently with several branches at or near the base; distinguished from woody vines in not requiring support from another plant or object.

SINUS.   The cleft or recess between two lobes of an expanded organ such as a leaf.

SPATULATE.   Like a spatula; somewhat widened toward a rounded end.

SPIKE.   A type of inflorescence in which stalkless flowers are attached along the sides of an elongated common axis.

SPINE.   A sharp-pointed rigid modified leaf or part of a leaf, as in Black Locust. See PRICKLE, THORN.

STAMEN.   The pollen-producing organ of a flower, usually consisting of anther and filament.

STELLATE.   Star-shaped. Applicable where several similar parts spread out from a common center, as with clustered hairs.

STIGMA.   The pollen-receptive part of a pistil, often enlarged, usually at the tip of the style.

STIPE.   Stalk of a pistil; not the pedicel. Stalk under elevated glands.

STIPITATE.   Having or borne on a stipe.

STIPULES.   A pair of structures, usually small, on the base of the petiole or on the stem near the petiole or on both; sometimes fused together.

STONE.   A seed with a bony covering, as in a Peach or Plum; a pit.

STYLE.   That portion of the pistil between stigma and ovary. Sometimes missing.

SUB-.   A prefix signifying almost, nearly, less than completely, somewhat. Subglobose implies not quite globe-shaped.

SUPERPOSED.   Attached above another part, i.e., toward tip from the first.

TERETE.   Circular in cross section.

THORN.   A hard sharp-pointed stem, as in Honey Locust and Hawthorn. See PRICKLE, SPINE.

TOMENTOSE.   Densely covered with soft fine matted woolly relatively short hairs. See VILLOUS.

TREE.   Plant with a perennial trunk a minimum of 4 m (13 feet) in height and a diameter at least 7.5 cm (3 inches) at breast height, namely 1.37 m (4.5 feet).

TRUNCATE.   An apex or base nearly or quite straight across, as if cut off.

TURBINATE.   Top-shaped; a solid having a tapering base and a broad rounded apex.

TWO-RANKED.   The attachment of alternate or opposite leaves on a stem in two opposite vertical rows, thus the places of attachment lying in one plane.

UMBEL.   A type of inflorescence in which flower stalks of approximately equal length arise from the same level on the stem like ribs of an umbrella. A compound umbel has a second group of umbels set on the first and sometimes a third group on the second.

VILLOUS.   Densely covered with soft fine unmatted relatively long hairs. See TOMENTOSE.

WHORL.   Three or more structures (leaves, stems, etc.) in a circle, not spiralled; e.g., three or more leaves at a node.

Key to Groups

1. Leaves narrow, under 4.5 mm wide; often needlelike, awllike, or scalelike.
   CONIFERS and *Tamarix*   Group A
1. Leaves with blades over 4.5 mm wide; blades linear to nearly orbicular.
   FLOWERING PLANTS   2
   2. Leaves with all veins parallel   Group B
   2. Leaves with netted veins, in some species some veins parallel to each other but
      not to all   3
      3. Leaves compound   4
         4. Leaves opposite   Group C
         4. Leaves alternate   Group D
      3. Leaves simple   5
         5. Leaves opposite or whorled   Group E
         5. Leaves alternate (some may be opposite in *Broussonetia*)   6
            6. Stipules and stipule scars encircling twig or nearly so   Group F
            6. Stipules and stipule scars absent or, if present, not encircling twig   7
               7. Leaves palmately veined   Group G
               7. Leaves pinnately veined   8
                  8. Leaves evergreen   Group H
                  8. Leaves deciduous or persisting and dead   9
                     9. Leaves, leaf scars, and axillary buds 2-ranked (check on
                        longest most vigorous growth for one year)   Group I
                     9. Leaves, leaf scars, and axillary buds in 3 or more ranks   10
                        10. Leaf margin entire   Group J
                        10. Leaf margin serrate to dentate   Group K

Conifers

Plants without flowers or fruits. Pollen, which develops in small cones, is airborne and falls directly on naked ovules that develop into seeds. Ovules and seeds are borne on the surface of scales arranged in cones, or sometimes on end of stalks.

Group A

Leaves narrow, under 4.5 mm wide, often needlelike, awllike, or scalelike.

KEY TO GENERA
1. Leaves opposite or whorled, not in tight clusters, evergreen　　　　2
　　2. Leaves linear, with a distinct petiolelike constriction at base, and with a fine
　　　　bristle tip　　　　　　　　　　　　　　　　　　　　　　　　　1. *Torreya*
　　2. Leaves scalelike to awllike or nearly linear, lacking a petiolelike base, acute to
　　　　sharp pointed but without a bristle tip　　　　　　　　　　　　　　3
　　　　3. Clusters of small branchlets bushy, i.e., three-dimensional; seed cones
　　　　　　berrylike, dropping without splitting open, seeds wingless　　13. *Juniperus*
　　　　3. Clusters of small branchlets essentially in one plane, as if pressed; seed
　　　　　　cones leathery or woody, splitting open at maturity and dropping later,
　　　　　　seeds winged　　　　　　　　　　　　　　　　　　　　　　4
　　　　　　4. Mid-portion of ultimate branchlets, including clasping part of leaves,
　　　　　　　　about 1.5 mm broad; seed cones oblong　　　　　　　11. *Thuja*
　　　　　　4. Mid-portion of ultimate branchlets, including clasping part of leaves,
　　　　　　　　about 1.0 mm broad; seed cones globose　　　　12. *Chamaecyparis*
1. Leaves alternate, distinctly so on vigorous twigs, usually scattered, sometimes in
tight clusters surrounded at base by a sheath, or some densely clustered without
sheaths on stubby twigs, deciduous or evergreen (*Taxus* may have some opposite
leaves also)　　　　　　　　　　　　　　　　　　　　　　　　　　5
　　5. Leaves under 4 mm long, plants with flowers, seeds with tufts of hairs (a
　　　　flowering plant)　　　　　　　　　　　　　　　　　　　14. *Tamarix*
　　5. Leaves over 4 mm long, plants without flowers or tufted hairs on seeds　6
　　　　6. All leaves in tight clusters of 2–5, those in each cluster parallel or nearly
　　　　　　so, at least toward their bases, surrounded at base by a sheath of scale leaves
　　　　　　(sheath falling early in *P. strobus*)　　　　　　　　　　3. *Pinus*
　　　　6. All leaves scattered (borne singly), OR only some scattered (on long twigs)
　　　　　　and others densely clustered (on stubby twigs), divergent and not sheathed
　　　　　　at base　　　　　　　　　　　　　　　　　　　　　　　　7
　　　　　　7. Some leaves scattered on leading twigs, others densely clustered on
　　　　　　　　stubby twigs　　　　　　　　　　　　　　　　　　　　8
　　　　　　　　8. Leaves evergreen, 4-angled in cross-section　　　　4. *Cedrus*
　　　　　　　　8. Leaves deciduous, 3-angled in cross-section　　　　5. *Larix*
　　　　　　7. All leaves scattered, although sometimes closely so on the shortest slow-
　　　　　　　　growing branchlets　　　　　　　　　　　　　　　　　9
　　　　　　　　9. Leaves deciduous, linear and flat to awl-shaped, tapering toward tip,
　　　　　　　　　　sharp-pointed　　　　　　　　　　　　　　　10. *Taxodium*

9. Leaves deciduous or evergreen, linear, sharp-pointed to blunt   10
  10. Leaves keeled above and below and thus 4-sided   6. *Picea*
  10. Leaves flattened, those of some genera with a prominent midrib
          11
    11. Leaves soft and flexible, with an abruptly acuminate tip that
       is not sharp and piercing to the touch; seeds borne singly
       and partially enclosed in a fleshy cup   2. *Taxus*
    11. Leaves soft and flexible to stiff, with notched to obtuse to
       acute tip (those of *Cunninghamia* sharp to touch); seeds
       several, borne in cones formed of spirals of several to many
       scales   12
      12. Leaves with very short petiolelike base   7. *Tsuga*
      12. Leaves obviously sessile   13
        13. Leaf margin finely serrate (rare escape)
                    9. *Cunninghamia*
        13. Leaf margin entire   14
      14. Leaves with two white bands below, evergreen; seed
        cones erect   8. *Abies*
      14. Leaves lacking white bands, mostly falling with
        deciduous branchlets, the others deciduous from
        persistent stems; seed cones not erect   10. *Taxodium*

TAXACEAE: Yew Family

1.
TORREYA. Torreya

Florida Torreya;
Stinking-cedar
Torreya taxifolia Arn.
[1]

Recognized by leaves linear, opposite, 25–40 mm long. Trees rarely over 10 m tall, in cultivation to 13 m tall by 87 cm DBH; parts with a strong disagreeable odor when crushed. Leaves stiff, with a short petiolelike constriction, often slightly curved, appearing 2-ranked but attached otherwise, apex with a long sharp point that is piercing to the touch. Pollen and seeds borne on separate trees. Seeds completely surrounded by a fleshy cup (aril) that at maturity is dark green with purple stripes, the whole drupelike and about 3 cm long; maturing in late summer of the second season. A fungal disease of the stems has destroyed many trees and may eradicate the natural populations; specimens growing in cultivation in scattered localities possibly may escape the disease. Plants tolerant of cold climate; a planted colony at Highlands, NC, is producing seeds. Rare. Wooded slopes and ravines along e side of Apalachicola River in Gadsden,

Liberty, and Jackson Cos., Fla, and in Decatur Co., Ga, where most of it has been eradicated in construction of Lake Seminole. Another species, *T. californica* Torr., is of rare and local occurrence in Cal. Mar–Apr.

2.
TAXUS. Yew

Florida Yew
Taxus floridana Nutt. ex
Chapm.
[2]

Recognized by leaves evergreen, flat, mostly alternate, 18–25 mm long, with a very short petiole and an abruptly acuminate apex. Shrubs or trees to 8 m tall. Leaves appearing 2-ranked but attached otherwise, apex sharp but not piercing to touch. Pollen and seeds borne on separate trees. Seeds ovoid, brownish, 5–7 mm long, set deep in a light-red fleshy cup (aril) about 10 mm broad; maturing in Sept or early Oct of first season. Seeds and fresh foliage poisonous if eaten by people or livestock. Photograph is of a cultivated *Taxus,* the species undetermined. Rare. Wooded slopes and ravines along e side of Apalachicola River in Gadsden and Liberty Cos., Fla; also a small colony in a *Chamaecyparis* swamp southeast of Bristol, Fla. The only other species of this genus in the SE is a shrub, *T. canadensis* Marsh., mostly of n distribution ranging southward locally to eTenn, nwNC, and WVa. Another tree species occurs in wUS.

PINACEAE: Pine Family

3.
PINUS. Pine

Pines are easily recognized by their evergreen needle-shaped foliage leaves in bundles of 2–5 surrounded at the base by a sheath of overlapping bractlike leaves (sheaths falling early the first season in White Pine). Needle-shaped leaves of all other genera lack this sheath. Each bundle of leaves is on a minute branch arising from the axil of a non-green bract; these bracts are the primary leaves. Trees, often with whorled branches. Pollen, which has two wings, borne in the spring in short-lived male cones. Seeds winged, borne in woody female cones that mature in autumn of the second year. Male and female cones on the same tree, both with many spirally arranged overlapping scales. Each of the scales of a female cone bears two seeds near the base and usually a prickle on the tip. Pines are a very important source of lumber, wood for paper pulp, and resin for turpentine, rosin, and other chemicals.

Seeds are eaten by birds; rodents eat seeds, bark, and leaves; deer browse on twigs and foliage. Vegetative parts may be poisonous to livestock and deer when eaten in large quantities. Thirteen species occur in the SE in contrast to 11 in the NE and 23 in wUS. The 13 pines of the SE fall naturally into two groups: SOFT PINES, having 5 needles enclosed at base by a sheath that falls early, with one vascular bundle in cross-section of each needle, and with wood soft and close-grained. White Pine is the only species in this group. HARD PINES, having 2–3 needles enclosed at base by a sheath that persists with the leaves, with 2 vascular bundles in cross-section, and with wood usually hard and close-grained. The other 12 species belong in this group. Hybrids can be expected between SE hard pines when growing in the same locality and shedding pollen simultaneously.

KEY TO PINUS SPECIES
1. Needles 5 in a bundle   1. *P. strobus*
1. Needles 2–3, uncommonly 4, in a bundle   2
  2. Primary non-green bractlike leaves and bud scales prominently fimbriate; evergreen needle leaves 20–45 cm long, in bundles of 3, rarely some 4 on the same tree   2. *P. palustris*
  2. Primary non-green bractlike leaves entire or edged with hairs but not fimbriate; evergreen needle leaves 3–25 cm long, in bundles of 2–4   3
    3. Needles in bundles of 3, or 2 and 3 on same tree, or rarely 3 and 4 on same tree   4
      4. Needles in bundles of 2 and 3 on same tree   5
        5. Needles 3–7 cm long, prickles on seed cones 3–8 mm long. Uncommon form of   6. *P. pungens*
        5. Needles 7–25 cm long, prickles on seed cones under 3 mm long. Common trees   6
          6. Needles 17–25 cm long; seed cones 6–15 cm long; no resin pockets on bark   3. *P. elliottii*
          6. Needles 7–13 cm long; seed cones 3–7 cm long; plates on bark with resin pockets, these sometimes scarce   5. *P. echinata*
      4. Needles in bundles of 3, or rarely 3 and 4 on same tree   7
        7. Cones sessile, longer than broad, opening at maturity and then falling soon; needles mostly 18–23 cm long, uncommonly as short as 10 or as long as 25 cm, 0.7–1.5 mm wide; stubby branches and/or clusters of needle bundles rarely present on trunk; buds not resinous   4. *P. taeda*
        7. Cones sessile or short-stalked, about as broad as long, nearly globose to ovoid when closed, nearly globose when open, opening at maturity or remaining closed and persistent for years (often up to 12); needles mostly 7–16 cm long, uncommonly as short as 4 or as long as 20 cm, 1.5–2.0 mm wide; stubby branches and/or clusters of needle bundles on trunk; buds resinous   8
          8. Needles 4–14 cm long, usually falling their second year; seed cones opening at maturity   11. *P. rigida*
          8. Needles 10–20 cm long, persisting 3–4 years; seed cones usually remaining closed for several years   12. *P. serotina*

3. All needles in bundles of 2   9
  9. Needles over 1 mm wide, mostly about 1.5 mm   10
    10. Seed cones 5–9 cm long, sessile, opening tardily, persisting for several years; spines stout, upwardly hooked, 3–8 mm long
        6. *P. pungens*
    10. Seed cones 3–6 cm long, stalked and usually reflexed, opening at maturity; spines absent (rare escape)   7. *P. sylvestris*
  9. Needles under 1 mm wide   11
    11. Needles 10–17 mm long, breaking with a sharp snap when bent; cones unarmed, in SE native only to WVa   13. *P. resinosa*
    11. Needles 2–10 mm long, bending without breaking or breaking tardily and without a sharp snap; seed cones armed (prickles minute and often few in *P. glabra*)   12
      12. Needles 2–8 cm long; seed cones opening at maturity, prickles prominent; plants of interior upland areas to the mid-Atlantic coast   8. *P. virginiana*
      12. Needles 5–13 cm long; seed cones opening at maturity or remaining closed for many years, prickles prominent or minute; plants of the southern CP   13
        13. Tip of seed cone scales with a conspicuous horizontal ridge, prickles short and stout; plants of well-drained white sands   10. *P. clausa*
        13. Tip of seed cone scales with horizontal ridge faint or missing, prickles minute; plants of fertile soils, usually lowland   9. *P. glabra*

1.
Eastern White Pine
Pinus strobus L.
[3]

Recognized by leaves in bundles of 5. One of the largest pines in the SE, fast-growing, once attaining heights of over 60 m and DBH around 1.8 m, now rarely half that size. Conspicuous whorls of nearly horizontal or ascending branches, one whorl produced each year, are characteristic of this species and useful in estimating the age of a tree. Needles 6–14 cm long, margin minutely serrate, persisting 3–4 years, the basal sheath usually shedding around midsummer. Mature cones 10–20 cm long, usually falling the first winter or the following spring, the scales thin, rounded, without prickles. Likely to be damaged by white pine blister rust if currants or gooseberry plants, which provide infective spores, are nearby. Wood soft, white, easily worked, once widely used, now scarce. Popular as an ornamental, doing best to the north and in the mountains; often pruned to desired size and shape. Needles and seeds an important food for birds and mammals. Common. Occupying a variety of habitats, wet to dry, sunny to shady, but does best in well-drained soils. Occurs to over 1500 m elevation in sAppalachians and down to sea level in the north. Extends to Nfld. A variety occurs in mountains of sMex and Guatemala. Seven other soft pines occur in wUS; one, *P. monticola* Doug. ex D. Don, Western White Pine, is scarcely distinguishable from *P. strobus*. Apr.

2.
Longleaf Pine
Pinus palustris Mill.
[4]

Easily recognized by having the longest needles (20–45 cm); by fimbriate bud scales and bracts subtending the bundles of needles. Mature trees rarely over 35 m tall or 1 m DBH. Seedlings usually with scant aboveground stem the first 3–6 years, sometimes longer, the dense bunch of needles formed easily mistaken for grass. Young stems, before the needles develop, resembling white candles; buds silvery-white in contrast to the brownish buds of other pines. Seed cones 15–25 cm long, scales with a small reflexed prickle. Resistant to fusiform rust, a serious disease of pines. Highly valuable for lumber, resin products, and pulpwood. Common. Grows best in open dry habitats; often associated with scrub oaks; does occur in flatwoods that are usually very wet in spring but exceedingly dry during summer. The most resistant to fire of SE species. Rare above 200 m elevation. Feb–Apr.

3.
Slash Pine
Pinus elliottii Engelm.
[5]

Recognized by relatively long needles (17–25, rarely 13–28 cm) in bundles of 2–3; by seed cones 6–15 cm long with pedicels 20–25 mm long. Rapid-growing trees to 45 m tall by 1 m DBH. Unlike *P. palustris,* terminal buds are rusty-silvery and bud scales and bracts not fimbriate. Young female cones are pointed forward at first but most of them soon curve sharply backward, remaining so until dropping. This character is useful in distinguishing the similar *P. taeda,* which has sessile cones that are commonly perpendicular to the stem. Scales of seed cones with short stout prickles. Fairly susceptible to fusiform rust. Abundantly planted in reforestation; important for lumber, pulpwood, and resin products, yielding more resin than Longleaf Pine. Common. Naturally occurring mostly in wet habitats in the open or moderate shade. Relatively susceptible to fire damage but plentiful on upland sites under modern fire-protection practices. Grows naturally to about 150 m elevation. Jan–Mar.

4.
Loblolly Pine
Pinus taeda L.
[6]

5.
Shortleaf Pine
Pinus echinata Mill.
[7, 8]

Identified by medium-sized needles 10–25 cm long, tapering at the tip, in bundles of 3; by cones 6–15 cm long, sessile, opening at maturity and falling soon, with the lower part of the prickles broad. Trees mostly to 35 m tall by 1 m DBH, but recorded to 55 m by 1.5 m. Needle and cone sizes of the common *P. echinata* overlap a little but both are usually shorter and the needles are in bundles of 2 and 3 on the same tree. A rapidly growing tree, spreading abundantly, and extensively used in reforestation. Perhaps the major source of wood for lumber and pulp. Often infected with a rust which produces prominent enlarged areas on twigs, branches, and trunks and releases quantities of yellow spores in the spring. Oddly, these spores infect only oaks, spores produced on the latter infecting the pines. Often damaged and sometimes killed by southern pine beetles. Common. In a wide range of habitats; a frequent occupant of abandoned fields; grows best in deep moist soils. Somewhat shade tolerant; mildly resistant to damage by fire. Rare above 600 m elevation. Feb–Apr.

Distinguished by relatively short needles (7–13 cm) in bundles of 2 and 3 on the same tree; by short-stalked cones 3–7 cm long, armed with prickles about 1 mm long. Trees to 39 m tall by 1.1 m DBH. Trunk of larger trees with flat broad reddish-brown plates that are generally somewhat rectangular; freshly exposed plate surfaces usually with resin pockets. Bark of 3–4-year-old twigs flaky as well as rough, an important character in separating from *P. virginiana, P. glabra,* and *P. clausa,* whose twigs have rough but non-flaking bark. Also see *P. taeda* for comparison with *P. echinata.* Cones usually releasing seed at maturity, and then remaining on the tree for many years. Unusual among pines in sprouting from the base when seedlings and young trees are cut down or tops killed by fire. Sometimes killed by southern pine beetles; resistant to fusiform rust. Wood used for lumber and pulp; seeds an important source of food for wildlife. Common. Usually in upland soils; frequent in old fields, often in pure stands. Fairly shade tolerant. Rarely occurs over 1000 m elevation. Feb–Apr.

6.
Table Mountain Pine
Pinus pungens Lamb.
[9]

Recognized by needles 3–7 cm long, about 1.5 mm wide, in bundles of 2 (rarely some 3 on the same tree); by sessile cones, usually abundant, 5–7 cm long, with prominent upwardly curving prickles 3–8 mm long, generally opening partly at maturity and persisting for many years. Slow-growing trees, rarely over 20 m tall by 70 cm DBH. Twigs and branches quite tough. A minor source of wood. Occasional. Dry slopes and ridges of mountains, especially in rocky areas, mostly between 900 and 1350 m elevation, but reported to 1750 m. Rare in the lower elevations of the Piedmont. May.

7.
Scots Pine
Pinus sylvestris L.

About the same size as *P. pungens* and having similar needles, 4–7 cm long, about 1.5 mm wide, and in bundles of 2. May be separated by the seed cones, which are usually shorter (3–6 cm), stalked, and without prickles. To 23 m tall by 1.4 m DBH. Rare. Planted, persisting, and sometimes spreading. Native of Eurasia.

8.
Virginia Pine
Pinus virginiana Mill.
[10]

Distinguished by bark of 3-year-old twigs and small branches rough but not flaking; by needles 2–8 cm long, 1 mm or less wide, in bundles of 2; by cones 4–7 cm long, opening at maturity but persisting for at least several years, bearing prominent slender prickles 2–4 mm long. Fairly slow-growing and short-lived trees rarely over 20 m tall by 70 cm DBH. *P. glabra* and *P. clausa* are similar but differ in distribution and in having smaller prickles on the cone scales, those of the former minute, those of the latter short and thickened. Seed cones on some plants of the latter remain closed for many years. Of little importance for lumber but used considerably for pulp; seeds an important source of food for wildlife. Common. Well-drained soils; often in pure stands, frequent in old fields and after fires. Not tolerant of over-topping by other trees. Usually occurs between 30 and 900 m elevation but reported to 1675 m. Mar–May.

9.
Spruce Pine
Pinus glabra Walt.
[11]

Recognized by 3–4-year-old twigs and small branches rough but not flaking; by needles 4–10 cm long, 1 mm wide or less, in bundles of 2; by seed cones 4–9 cm long, scales with minute prickles and a faint horizontal ridge, usually falling soon or some in a few years. Trees to 35 m tall by 1.2 m DBH. Bark unlike that of other pines, at first smooth and dark gray, later becoming closely ridged, producing small thin flat plates. These characteristics are suggestive of Spruce, thus its common name. Except for the trunk, much like *P. clausa* and *P. virginiana;* the former may also be recognized by having a conspicuous horizontal ridge on seed cone scales, the latter by having slender prickles 2–4 mm long on the seed cones. Produces good-quality lumber and pulp; excellent as an ornamental. Occasional. Trees rarely in groups. Rich woods and hammocks, bottomland woods of CP. Quite shade tolerant. Feb–Apr.

10.
Sand Pine
Pinus clausa (Chapm. ex Engelm.) Vasey ex Sarg.
[12]

Recognized by bark of 3-year-old twigs and small branches rough but not flaking; by needles in bundles of 2, under 1 mm wide, 5–9 cm long; by tip of seed cone scales with a conspicuous horizontal ridge. Trees rarely over 25 m tall by 1.0 m DBH, usually much smaller. Cones of two types: In wFla and adj. Ala the cones open at maturity. The photograph is of one of these pines. In peninsular Fla the cones remain unopened on the trees for many years, sometimes so long as to become embedded in the enlarged branches and even the main stem. If trees are killed by a crown fire, the cones open. The first form is difficult to separate from the similar *P. virginiana* and *P. glabra,* whose cones open at maturity, except by distribution, but the second form is easily separated by cones remaining closed for many years. For other differences see these two species. Often in pure stands. Used some for pulpwood. Common locally. In open in white sands of ridges and hills. Planted locally; reproducing naturally in sGa. Feb–Mar.

11.
Pitch Pine
Pinus rigida Mill.
[13]

Recognized by needles being short (4–14 cm), about 2 mm wide, in bundles of 3 or rarely some 4 on same tree, abruptly sharp-pointed, and usually falling the second year; by seed cones 3–7 cm long, with slender sharp prickles. Trees rarely over 25 m tall by 1 m DBH. An unusual feature is the presence of stubby branches and/or clusters of needle bundles on the trunk. Cones nearly sessile, ovoid to nearly globose when closed, mostly globose when opened, often persisting unopened for many years, then opening after a fire. The similar *P. serotina* is separated by having usually longer needles (12–20 cm) which persist for 3–4 years and have long tapering ends, and by cones that usually remain closed for years. Unusual among pines in sprouting from the base if seedlings or young trees are cut down or tops killed by fire. Resistant to decay. Used for lumber and pulpwood; once used as a source of resin; twigs, leaves, and seeds an important source of food for wildlife. Common. Poor sandy soils and a variety of other habitats but intolerant of competition with other trees. Fire resistant. From about 300 m to over 1700 m elevation in the southern mountains, down to sea level along the nAtlantic. May.

12.
Pond Pine
Pinus serotina Michx.
[14]

Recognized by needles 10–20 cm long, 1.5–2.0 mm wide, with long tapering tips, in bundles of 3 or sometimes 4 on the same tree, falling in 3–4 years; by seed cones 3–6 cm long, subglobose to ovoid when closed, persistent and frequently remaining unopened for several years, stalked but later sometimes becoming sessile or imbedded as the branches enlarge, prickles small and often dropping. Trees to 25 m tall by 0.8 m DBH; crown often open and irregular. Usually with stubby branches and/or clusters of needle bundles on the trunk, and sprouting from the base and roots after fires or other damage, both unusual features in pines. Similar in several respects to *P. rigida*, which see for comparative features; the two species hybridize where ranges overlap. Valued mostly for pulpwood. Common. Swamps, ponds, poorly drained areas that usually have striking changes in groundwater levels and are subject to frequent fires during dry periods. Mar–Apr.

13.
Red Pine; Norway Pine
Pinus resinosa Ait.
[15]

Identified by slender needles under 1 mm wide, 10–17 cm long, in bundles of 2, persisting for 4–5 years, and breaking in half when doubled between fingers. Trees rarely over 25 m tall by 1 m DBH, but recorded over 38 m by 1.5 m. Young cones near tip of twigs; seed cones ovoid-conic when closed, 4–6 cm long, with no prickles, opening and shedding soon after maturing. Source of wood for lumber and pulp; used as an ornamental; seeds valuable food for wildlife. Rare in SE, occurring only in WVa, but common to the north. Usually in well-drained soil. Shade tolerant, competing well with other trees as well as growing in the open. Often in pure stands. Occurs around 1150 to 1300 m elevation in WVa but much lower northward. Extends to Nfld. May.

4.
CEDRUS. Cedar

Deodar Cedar
Cedrus deodara Loud.
[16]

Recognized by tips of branches drooping; leaves evergreen, alternate, 2.5–5 cm long, 4-angled in cross-section, with many clustered on lateral spurs, those on the terminal twigs of the year scattered. Trees to 45 m tall. Seed cones 9–13 cm long, dropping the scales the second year, leaving the cone axis attached. Seeds with a large wing. Frequently planted as an ornamental, only occasionally reproducing naturally from seeds. Native of Himalayas.

Cedar-of-Lebanon
Cedrus libani Loud.

Similar to *C. deodara* but may be separated by having horizontal branch tips. It is less frequently planted and we have no record of natural reproduction, but this occurrence is logical. Native of Asia Minor.

5.
LARIX. Tamarack; Larch

Recognized by leaves deciduous, very slender, 3-angled, over 15 mm long, obviously alternate on leading twigs but not obviously alternate on the dwarf lateral branches that grow only enough each year to develop the dense cluster of leaves. Male and female cones on separate trees. Seed cones erect, the scales nearly orbicular; maturing within a year and shedding seeds, then usually remaining on tree for another year. Seeds winged. Two species occur in SE, 2 in nwUS and swCan.

KEY TO LARIX SPECIES
1. Seed cone scales 10–20, glabrous and shiny outside   1. *L. laricina*
1. Seed cone scales 30–50, finely to minutely hairy outside.   2. *L. decidua*

1.
Tamarack; Eastern Larch
Larix laricina (Du Roi) K. Koch
[17]

Recognized by seed cones 12–20 mm long, on dwarf branches of a previous year's growth, scales 10–20. Trees to 25 m tall by 0.6 m DBH, rarely 30 m by 1 m, usually slow-growing. Leaves 2–3 cm long, soft, the apex blunt. Used mostly for lumber and pulp; seed an important source of food for wildlife. Rare in SE but common northward. Wet soils of bogs and swamps in s part of range, often in better drained upland soils northward, doing best in drier habitats. Very susceptible to fire damage but usually affected only in the drier habitats. Occurs to ca. 1200 m in s part of range but down to near sea level northward. Extends to Alas, Ont, and Lab.

2.
European Larch
Larix decidua Mill.

Similar to *L. laricina* but cones larger, 20–35 mm long, and with over twice as many scales. Rare. Used sparingly in reforestation, extensively as an ornamental in the north; sometimes reproducing naturally. Native of Europe.

6.
PICEA. Spruce

Recognized by leaves alternate, sessile but raised on peglike twig projections 0.1–0.2 mm long, keeled above and below and thus 4-sided, sharp-pointed, the blades scattered around the twig somewhat like a bottlebrush. Large trees with whorled branches. Leafless twigs rough because of the persistent peglike leaf supports. Male and female cones borne on ends of previous year's growth. Seed cones pendulous, maturing the first year. Seeds winged. There are about 40 species in the cooler parts of the N. Hemisphere; 6 are native to the US and Can, and only one occurs in the SE. An exotic species, *P. abies,* is widely planted, escaping in the SE and elsewhere.

KEY TO PICEA SPECIES
1. Upper branches ascending, the lower spreading; outer bud scales prolonged into minute hairlike projections   1. *P. rubens*
1. Upper branches spreading to ascending, the lower drooping, especially at ends; outer bud scales without hairlike projections   2. *P. abies*

1.
Red Spruce
Picea rubens Sarg.
[18]

Identified by spreading lower branches; by leaves alternate, scattered, evergreen, 4-sided, sharp-pointed; by hairlike tips on outer bud scales. Trees to over 45 m tall by 1.3 m DBH; shallow-rooted and subject to uprooting by winds. Crown narrow and conical. Seed cones hanging down, 3–4 cm long, reddish-brown when mature. Often defoliated by the spruce budworm. Once an important source for lumber and pulpwood, but cutting has drastically reduced the supply. Twigs, leaves, and seeds an important source of food for wildlife. Common. Mostly in upland areas in pure stands or mixed with other species. Shade tolerant; highly susceptible to fire. From about 1060 to 2000 m elevation in sAppalachians, where extensive virgin stands occur, to near sea level in Canada. The similar but generally more northern *P. mariana* (Mill.) BSP., Black Spruce, has been reported in at least 3 SE states, apparently erroneously. It is distinguished by shorter seed cones, 1.5–3 cm long, that are dull gray when mature. May–June.

2.
Norway Spruce
Picea abies (L.) Karst.

Separated from native e spruces by drooping branches and longer seed cones, 10–18 cm long. Rare. Widely planted as an ornamental and in local reforestation in NE and Can; reproducing naturally. A native of Europe; with many horticultural varieties.

7.
TSUGA. Hemlock

Recognized by leaves evergreen, flattened, 2 white stripes beneath, apex obtuse, originating alternately on all sides of the twig, with short petiolelike stalks attached to short projections from the twig. After leaves fall the twigs are rough because of these projections. Medium to large trees. Leaf scars perpendicular to the twig. Seed cones pendulous, borne at end of previous year's lateral branches, maturing late the first year. Cone scales persistent, the cone falling intact long after seeds are shed. Seeds winged. Two species occur in wUS, wCan, and sAlas.

KEY TO TSUGA SPECIES
1. Leaves spirally arranged but the blades spreading in one plane, margins minutely toothed; seed cones brownish, 12–25 mm long   1. *T. canadensis*
1. Leaves spirally arranged and spreading in all directions; margins entire; seed cones yellowish, 20–38 mm long   2. *T. caroliniana*

1.
Eastern Hemlock
Tsuga canadensis (L.) Carr.
[19]

Recognized by "petioled" flattened obtuse leaves originating alternately on all sides of the stem but "petioles" bent so that blades are essentially in one plane; margin minutely toothed. Trees to 48 m tall by 1.9 m DBH with shallow root system. Seed cones 12–25 mm long, short-stalked, hanging, shedding seeds in autumn, persisting until the next year or longer. Bark once an important source of tannin for leather production. Wood of poor quality for lumber. Valuable tree as an ornamental in or near its natural range; occasionally trimmed into hedges. Seeds provide an important source of food for birds and rodents. Common. Does best in moist shaded situations, but occurs in dry open places. To 1740 m in sAppalachians but scarce above 1500 m; down to sea level in the north. Mar–Apr.

2.
Carolina Hemlock
Tsuga caroliniana Engelm.

Similar to but usually smaller than *T. canadensis*. It is distinguished by having leaf blades spreading from twig in all directions, the margin entire; by seed cones 20–38 mm long. Little value for wood products; prized as an ornamental. Occasional. Dry slopes, cliffs, bluffs, rocky ridges; occasionally in moist places. From 640 to 1200 m elevation. Mar–Apr.

8.
ABIES. Fir

Recognized by leaves sessile, evergreen, fragrant, usually curved, flat, with two white bands beneath, originating alternately on all sides of the stem. Medium to large trees. Fresh leaf scars flush with stem surface. Seed cones erect, the scales falling late the first year, leaving the axis standing. Seeds winged. Seven species occur in wUS and wCan.

KEY TO ABIES SPECIES

1. Bracts of seed cones strongly protruding beyond scales and turned downward
                                                                    1. *A. fraseri*
1. Bracts of seed cones hidden by the scales or protruding slightly and not turned
   downward   2. *A. balsamea*

1.
Fraser Fir; She-balsam
Abies fraseri (Pursh) Poir.
[20]

Recognized by cones with strongly protruding and downwardly turned bracts that essentially hide the scales. Trees to 25 m tall by 79 cm DBH; shallow-rooted, subject to windthrow. Foliage soft to touch, with a strong resinous odor when rubbed. The balsam woolly aphid, which was introduced into nUS from Europe in 1908, was discovered on *A. fraseri* in 1963. The aphid has been responsible for the death of most of the trees and may eradicate the species. Unprotected stands of timber are too scarce for this species to be of much commercial value; popular as a Christmas tree. Common at high elevations in mountains from about 1200 to 2000 m; a component of the extensive spruce-fir forests of the Appalachian Mts. Susceptible to damage by fire. About 150 trees were planted between 1957 and 1959 on Brasstown Bald in nGa; 25–30 trees were still living in 1983. May–June.

2.
Balsam Fir;
Canada Balsam
Abies balsamea (L.) Mill.
[21]

Recognized by having bracts of seed cones hidden by the scales or only slightly protruding. Trees to 38 m tall by 0.9 m DBH, but usually to only 20 m by 0.5 m; at timberline often dwarfed and matted. Leaves on the upper branches usually strongly curved upward; those on the lower usually not. Susceptible to severe damage by windthrow, fire, spruce budworm, and the balsam woolly aphid. Interestingly, Balsam Fir contains a complex chemical that is a growth regulator for many insects. A major source of pulpwood; popular as a Christmas tree. Canada balsam, another product, is a liquid used to cover thin specimens for permanent mounting under cover glasses on glass slides for microscopic examination. Foliage and seeds important as food for wildlife. Vegetatively very similar to *A. fraseri* but their natural ranges are well separated. Rare in SE, common in north. Usually in moist places such as swamps. Above 1200 m in south, in north from sea level to timberline; in pure stands or with other conifers, birch, or aspen. Extends to Lab and Alta. May–June.

TAXODIACEAE: Redwood Family

9.
CUNNINGHAMIA. China-fir

China-fir
Cunninghamia lanceolata
(Lamb.) Hook.
[22]

Identified by leaves evergreen, flattened, rigid, quite sharp-pointed to touch, fastened alternately all around twig, bases broad and decurrent, margin finely serrate, with 2 broad whitish bands beneath. Trees to 25 m tall with whorled branches. Seed cones globose-ovoid with leathery serrate sharp-pointed scales. Seeds winged. Occasionally planted as an ornamental, reproducing naturally in a few localities. Native of China. Syn: *C. sinensis* R. Br.

10.
TAXODIUM. Bald-cypress

Recognized by leaves deciduous, linear and flat to awl-shaped, fastened alternately all around twig, some on persistent terminal twigs, others on lateral branchlets most of which fall off in autumn with leaves attached. Small to large trees with grayish to reddish-brown bark, base of trunks enlarged and usually spreading into rounded ridges or prominent buttresses. Male and female cones on same tree, young ones evident in the late summer of the year preceding pollination and full development of the cones. Mature seed cones globose or nearly so, resinous, the scales peltate, cone disintegrating when mature. Seeds winged. Roots usually develop erect conical "knees." Another species in extreme s tip of Tex and in Mex. Not true cypresses, which are of the genus *Cupressus,* none occurring naturally in the SE.

KEY TO TAXODIUM SPECIES
1. Branchlets spreading laterally from the larger twigs; leaves linear, flat, mostly spreading in one plane (appressed only on drooping branches of the crown, if at all)   1. *T. distichum*
1. Branchlets mostly rigidly ascending; leaves slender, needle-shaped to awl-shaped, spirally arranged on and appressed to the branchlets (spreading in one plane only on lower limbs, sprouts, or saplings, if at all)   2. *T. ascendens*

A

1.
Bald-cypress
Taxodium distichum (L.)
L. C. Rich.
[23]

Recognized by flat linear leaves with blades mostly spreading in one plane; by bark that is shallowly furrowed, ridged on older trunks, and sloughing in thin flaky scales. Trees often to 30 m tall by 1 m DBH, recorded to 45 m by 3.8 m. Heartwood highly resistant to decay, thus practical for uses involving contact with soil or exposure to weather. Highly prized as lumber for interior trim. Excellent as an ornamental, growing rapidly in deep moist soils in open. Hybridizes with *T. ascendens,* especially in w and ne parts of the latter's range. Common. Live bark easily killed by fire but uncommon due to the habitat. Swamps, floodplains, river banks, sometimes in brackish tidewater. Hardy north into Can. Mar–Apr. Syn: *T. d.* var. *distichum.*

2.
Pond-cypress
Taxodium ascendens
Brongn.
[24]

Recognized by slender needle-shaped to awl-shaped leaves appressed to the stems, spreading to one plane on sprouts and saplings and occasionally on lower branches; by bark prominently ridged and deeply furrowed and thicker than in *T. distichum*. Trees to 41 m tall by 2.3 m DBH. Stems of Climbing Heath, *Pieris phillyreifolia* (Hook.) DC, frequently grow upward out of sight inside the dead bark, sending lateral branches out from the trunks, giving Pond-cypress the appearance of having wide as well as linear leaves. Wood of quality equal to that of *T. distichum*. Because of hybridization and intergradation with the latter species, *T. ascendens* is considered by some a variety, *T. distichum* var. *imbricarium* (Nutt.) Croom. or *T. d.* var. *nutans* (Ait.) Sweet. Preliminary results of planned extensive studies favor treating them as varieties. Common. Shallow ponds, lake margins, wet sandy depressions, swamps. Mar–Apr.

CUPRESSACEAE: Cypress Family

11.
THUJA. Arborvitae

Northern White-cedar;
Eastern Arborvitae
Thuja occidentalis L.
[25]

Recognized by opposite scalelike evergreen leaves; by clusters of branchlets in one plane, the ultimate branchlets about 1.5 mm broad. Slow-growing trees, unusual if over 20 m tall by 0.8 m DBH, but recorded

to 38 m by nearly 2 m. Leaves 3–7 mm long, the side pair keeled, the other pair flat and each with a gland-dot. Seed cones cylindrical about 7–12 mm long, with 3–6 pairs of overlapping scales, maturing and opening the first year. Seeds with a small wing. The wood highly resistant to decay; has moderate use for poles, cross-ties, and lumber. Popular as an ornamental. Resembles *Chamaecyparis thyoides* which may be distinguished by globose cones about 6 mm in diameter and mid-portion of ultimate branchlets about 1.0 mm broad including clasping part of leaves. *Juniperus* species are also similar, but clusters of small branchlets spread in many directions and cones drop without opening. Common. Wet peaty acid soils; peculiarly also on limestone outcrops. Moderately shade tolerant. To cMan and Anticosti Is. Another species occurs in nwUS and swCan. Mar–Apr.

12.
CHAMAECYPARIS. White-cedar

Atlantic White-cedar
Chamaecyparis thyoides (L.)
B.S.P.
[26]

Recognized by small evergreen opposite scalelike leaves; by flattened cross-sections of smallest twigs; by clusters of small branchlets in one plane; the ultimate branchlets about 1 mm broad. Trees uncommon over 25 m tall and 1 m DBH, but recorded to 36 m by 1.5 m. Leaves 1.5–4 mm long. Male cones small, brown, persistent after shedding pollen early in year (note in photograph). Seed cones globose, bearing 6 peltate scales, usually remaining on the tree at least until new cones are fully developed the following year, as may be seen in the photograph. Seeds winged, shed in autumn. Wood of good quality for lumber but little available for harvesting. Hardy and attractive as an ornamental, having an advantage over Redcedar in being insusceptible to apple-cedar rust. See the similar *Thuja occidentalis* for differences with it. Easily confused with *Juniperus* species, but they have clusters of small branchlets spreading in all directions and cross-sections of smallest twigs are quadrangular. Occasional. Swamps, terraces along clear-water streams. Shade tolerant. Two species occur in the area from nCal to seAlas. Mar–Apr.

13.
JUNIPERUS. Juniper; Redcedar

Recognized by clusters of small branches spreading in several directions, cross-sections of smallest twigs quadrangular; by leaves opposite or whorled, scalelike, awllike, lancelike, or linear. Shrubs to medium-sized trees, evergreen, with fluted trunks. Male and female cones usually on different trees. Mature seed cones berry-like, dropping with seeds. Sometimes confused with *Chamaecyparis* and *Thuja* but they have clusters of small branchlets flattened and cross-sections of smallest twigs are flattened. Four species occur in the SE. Some of these and 7 others occur in ne and wUS, Alas, and Can.

KEY TO JUNIPERUS SPECIES
1. Leaves in whorls of 3; cones axillary, maturing in 2–3 years   1. *J. communis*
1. Leaves opposite, rarely whorled; cones terminal on short branches, maturing first season   2
  2. Leaves entire, gland elongated; seeds with numerous pits   2. *J. virginiana*
  2. Leaves minutely serrate (use at least 20 ×), gland on leaves circular; seeds lacking pits   3. *J. ashei*

1.
Common Juniper
Juniperus communis L.
[27]

Recognized by awl-shaped to linear leaves 5–20 mm long in whorls of 3. Evergreen shrubs or trees to 10 m tall by 14 cm DBH. Seed cones berrylike, axillary, maturing in 2–3 years. Note two ages of cones in photograph. Southernmost colonies usually shrubs, but reported as a tree as well as a shrub from Md northward. Useful ornamental. Seed cones important food for wildlife; produce ingredient for gin. Rare in SE, common northward. Well-drained soils, usually in open areas. Circumboreal. Mar–Apr.

2.
Eastern Redcedar
Juniperus virginiana L.
[28]

A variable species. Mature trees recognized by leaves opposite, entire, ovate to triangular, 1.5–3 mm long. On very young trees some leaves may be whorled. Plants up to 3 m tall and sometimes the lower branches of larger trees often with leaves 3-sided, 4–7 mm long, linear, spreading, and sharp-pointed. Also recognized by erect fleshy seed cones on short leafy twigs, maturing the first year; by seeds with numerous pits. Trees to 35 m tall and 1.2 m DBH, but usually much smaller. Crown columnar to broadly cone-shaped

to rounded. Bark exfoliating in narrow strips. Twigs often developing hard swollen segments due to infection by a rust fungus that produces conspicuous gelatinous masses containing orange spores in the spring. These spores infect wild and cultivated apple trees, often causing considerable damage. In turn, spores produced on the apple trees infect the cedars. Wood highly resistant to decay, much used as fenceposts. Lumber used for chests, cabinetwork, paneling, carvings. Seed cones valuable food for wildlife. We consider *J. virginiana* as consisting of two poorly defined varieties, *virginiana* and *silicicola* (Small) Siba (Southern Redcedar). Population studies, including chemical analyses by Robert P. Adams, published in February 1986, support this contention. Seed cone diameters in the species vary from 5 to 7 mm long, width of terminal twigs from 0.75 to 1.00 mm, and scale leaf lengths from 1.20 to 1.65 mm, with var. *silicicola* tending to have the smaller sizes and var. *virginiana* the larger. Var. *silicicola* also tends to have a rounded crown, var. *virginiana* a crown tapered toward the top. The former variety occurs in Fla and extends along the coasts to Miss and NC, growing in thin woods or in the open, including established dunes and brackish sites; the latter variety has an inland distribution. Not a true Cedar, which is the genus *Cedrus*. Common. Usually well-drained soils in the open, rock outcrops, stable dunes. Not shade tolerant. Absent from high elevations in sAppalachians. West to cTex and swND. Jan–Apr.

3.
Ashe Juniper
Juniperus ashei Buchh.

Similar to *J. virginiana,* but leaves minutely serrate instead of entire, gland on leaves circular instead of elongated, or sometimes missing, and seeds lacking pits. Shrubs or trees to 11 m tall by 90 cm DBH. Crown rounded to irregular or open. Leaves mostly 1–2 mm long. Susceptible to cedar-apple rust. Uses similar to *J. virginiana*. Rare in SE. Limestone outcrops. To c and csTex, and adj. Mex. Feb–Mar. Syn: *J. mexicana* Spreng.

Flowering Plants

Plants with flowers and fruits. Pollen, developed in anthers, falls on the stigma of pistils, where it germinates, produces a pollen tube that penetrates the tissues of the pistil, and progresses to the ovule borne within. The ovules then develop into seeds which are enclosed in fruits formed from the pistil and sometimes some associated parts.

Group A

TAMARICACEAE: Tamarisk Family

14.
TAMARIX. Tamarisk

Recognized by leaves alternate, awllike or lance-shaped, clasping or sheathing the twig, under 3 mm long, usually about 1.5 mm, sometimes persistent through mild winters. Shrubs or trees to 10 m tall, with irregularly and widely spreading slender and limber twigs. Many of the branchlets fall with the leaves in autumn. Flowering randomly throughout the year; especially attractive when in flower. Flowers about 2 mm across, arranged in dense racemes; sepals and petals of same number, 4 or 5 each, 1–8 mm long. Fruits capsules about 5 mm long, releasing tiny seeds, each with a tuft of hairs that effectively disperse the seeds. Resembles *Juniperus,* but leaves are alternate instead of opposite or whorled. Planted as ornamentals and to stabilize loose sandy soils, especially along the coast; escaping and on some islands of the Gulf, so abundant in places as to eliminate much of the native vegetation. We are following a worldwide study of *Tamarix* because current manuals covering various parts of the SE are of limited value in naming species. Identification of the 6 species reported as naturalized in the SE is generally difficult and may require as much as 20 × magnification to see the minute diagnostic characters. Natives of E. Hemisphere.

KEY TO TAMARIX SPECIES
1. Flowers with 4 sepals and 4 petals   6. *T. parviflora*
1. Flowers with 5 sepals and 5 petals   2
    2. Stamens fastened between lobes of the disc, lobes thus obvious between the stamens; petals persistent after maturity   3
        3. Sepals entire or nearly so, petals ovate to elliptic   5. *T. chinensis*
        3. Sepals toothed, petals obovate   4. *T. ramosissima*
    2. All stamens arising from the disc lobes, the lobes thus seemingly absent; petals dropping soon after maturity   4
        4. Racemes 6–9 mm broad, mostly on darkened previous year's branches
                                                                            3. *T. africana*
        4. Racemes 4–5 mm broad, mostly on green branches   5
            5. Rachis of racemes usually papillose, sepals densely and finely incised-toothed, petals obovate   1. *T. canariensis*
            5. Rachis of racemes glabrous, sepals entire or nearly so, petals elliptic to slightly ovate-elliptic   2. *T. gallica*

1.
Salt-cedar; Tamarisk
Tamarix canariensis Willd.
[29]

Identified by racemes 4–5 mm broad, 5 sepals, 5 obovate petals which drop soon after maturity. Recorded from only a few localities in the SE. In brackish as well as non-brackish habitats, usually in sandy soils.

2.
Salt-cedar; Tamarisk
Tamarix gallica L.

Much like *T. canariensis,* but petals elliptic to slightly ovate-elliptic and sepals entire or nearly so instead of densely and finely toothed as in the latter. Reported as the most common species by local floras of the SE, but considered as rare in the worldwide study of the genus.

3.
Tamarisk
Tamarix africana Poir.

Much like the previous two species, but racemes 6–9 mm broad and commonly on previous year's growth instead of mostly on green branches as in the former two species. Reported for SC only, but to be expected elsewhere.

4.
Tamarisk
Tamarix ramosissima
Ledeb.
[30]

Recognized by 5 sepals which are toothed and 5 obovate petals which persist long after maturity. Rare. Reported for Ark by the worldwide study.

5.
Tamarisk; Salt-cedar
Tamarix chinensis Lour.

Much like *T. ramosissima,* but sepals entire or nearly so and petals ovate to elliptic. Often misidentified as *T. gallica,* but can be easily separated since the petals drop soon after maturity in the latter. Rare, but perhaps the most frequently encountered in the SE. Syn: *T. juniperna* Bunge.

6.
Tamarisk
Tamarix parviflora DC.

The most easily recognized of the species reported for the SE, with its 4 sepals and 4 petals instead of 5 as in all other species. Rare. Not known from brackish habitats.

Group B

Leaves with all veins parallel.

ARECACEAE: Palm Family

1.
SABAL. Palmetto

Cabbage Palmetto
Sabal palmetto (Walt.)
Lodd. ex Schult.
[31]

Recognized by fan-shaped leaves and petioles lacking spines. Branchless trees to over 20 m tall. Trunk generally of uniform diameter from base to summit, the diameter remaining constant with age. Leaves persistent, palmately divided into many parallel-veined segments, with a prominent arching midrib. Leaf blades 1–2 m long, the segments filamentous on margin and usually at tip. Petioles about 2 m long. Flowers numerous in large much-branched drooping clusters longer than the leaves; sepals 3, about 2 mm long; petals 3, 3–4 mm long. Fruits shiny, black, dry, 6–12 mm diameter, with one brown seed. Seedlings unusual in that the stem grows downward deeper into the ground for several years before turning upward permanently. Trunks used for wharf pilings as they are not injured by sea-worms. Often used as an ornamental. Common. Dunes, hammocks, flatwoods, brackish areas; in shade or open. Not known to occur naturally over 75 miles from the coast. June–July.

    S. minor (Jacq.) Pers., the Bush Palmetto, attains tree size in Tex, to about 7 m tall, but not in the SE, reaching a maximum of 2.5 m in La. Petioles unarmed. Leaf blades similar to those of *S. palmetto,* but flat and lacking fibers on margins. Common. Low woods. Fla to sTex, seOkla, cGa, and seNC. Apr–June.

2.
SERENOA. Saw Palmetto

Saw Palmetto
Serenoa repens (Bartr.)
Small
[32]

Recognized by leaf blades fan-shaped and petioles armed with stiff spines. Shrubs with branched trailing stems to small occasionally branched trees. A specimen seen on Sapelo Is., Ga, was 7.7 m tall. Trunk of light porous wood, the diameter remaining constant with age. Leaf blades green or glaucous, to about 1 m across, the segments lacking marginal fibers; petiole ending abruptly in the blade base. Flowers many in large much-branched clusters arising from axils of leaves; sepals 3, 1–2 mm long; petals 3, 3–5 mm long; stamens 6. Fruits ellipsoid to subglobose, 15–20 mm across, orange-colored when mature. The photograph is of a tree on Cumberland Is., Ga, more conspicuous than usual because of a fire. Common as a shrub, rare as a tree. In numerous sunny to shady habitats except swamps and poorly drained river terraces, although tolerant of short term water-logged soils. Exceptionally resistant to fire. May–July.

AGAVACEAE: Agave Family

3.
YUCCA. Yucca

Recognized by leaves alternate, closely arranged, leathery, evergreen, linear or nearly so, tapering to a sharp point. Shrubs or small trees. Trunk of light fibrous wood, the diameter remaining relatively constant with age. Flowers bell-shaped, pendulous, in a large terminal panicle. Sepals and petals white, 3 each and alike, stamens 6, carpels 3 and united. Fruit oblong, with 6 rows of seeds. *Yucca* pollination is interesting in that a *Pronuba* moth stuffs pollen into the stigma, usually after laying eggs in the ovary.

KEY TO YUCCA SPECIES
1. Leaf margin entire, with a very narrow thin brownish edge   1. *Y. gloriosa*
1. Leaf margin with small very sharp spiny serrations, especially on the lower half; lacking brownish edge   2. *Y. aloifolia*

1.
Moundlily Yucca
Yucca gloriosa L.
[33, 34]

Recognized by crowded bayonetlike leaves with entire margins that have a very narrow thin brownish edge. Shrubs or trees to 5 m tall. Trunk usually without branches, any branches short and densely covered with leaves; often with dead leaves clinging below live ones. Flowers 4–5 cm long. Fruits 5–8 cm long; when mature hanging down, dry, leathery, not splitting open. Small plants may be confused with the shrubby beargrasses, *Y. filamentosa* L. and *Y. flaccida* Haw., but neither of these has the brownish leaf margins, instead usually bearing loose fibers. Occasional. Dunes and other open sandy areas along coast. Aug–Oct.

2.
Spanish-bayonet
Yucca aloifolia L.
[35]

Recognized by crowded bayonetlike leaves with tiny sharp spiny serrations on margin and a dark very sharp tip. To 6 m tall, occasionally with one to a few upright leafy branches. Leaves thick and stiff. Flowers 4–6 cm long. Fruits 7–10 cm long; when mature hanging down, fleshy, but finally becoming nearly dry. Often used as an ornamental but dangerous because of sharp-pointed leaves. Occasional. Coastal dunes and other sandy soils, usually in open. Also an escape from plantings made throughout much of the SE. Rarely found in mountains. June–July.

Group C

Leaves compound and opposite.

KEY TO GENERA

1. Leaves palmately compound, leaflets 5–11   2
   2. Leaflets entire, twigs finely hairy   5. *Vitex*
   2. Leaflets irregularly serrate, twigs glabrous   3. *Aesculus*
1. Leaves pinnately compound, leaflets 3–15   3
   3. Leaflets 3 and finely serrate   1. *Staphylea*
   3. Leaflets 3, with coarse teeth, and sometimes lobed; OR leaflets 5–14   4
      4. Year-old twigs with large lenticels, the pith half or more the diameter; fruits berrylike juicy drupes with 3 seedlike nutlets   6. *Sambucus*
      4. Year-old twigs with lenticels small or lacking, the pith less than half of the diameter; fruits dry   5
         5. Leaflets 3–7, at least some leaflets lobed; fruits double, with two papery wings   2. *Acer negundo*
         5. Leaflets 5–11, none lobed; fruits single, with one papery wing   4. *Fraxinus*

STAPHYLEACEAE: Bladdernut Family

1.
STAPHYLEA. Bladdernut

American Bladdernut
Staphylea trifolia L.
[36]

Recognized by opposite compound leaves with 3 finely serrated leaflets. Shrubs, or uncommonly trees to 8 m tall by 15 cm DBH. Terminal buds absent. Flowers somewhat cylindrical, in axillary drooping clusters; sepals separate, greenish-white, and fastened to a disc located below the ovary; petals 5 each, the sepals united at base. Fruits ellipsoid to subglobose, drooping, inflated, with 3 pointed lobes, 3-celled, containing a few seeds. Vegetatively somewhat like *Ptelea,* but *Ptelea* leaves are alternate and the leaflets usually entire. Occasional. Rich deciduous woods. Generally absent from the Appalachian Mts. Another species occurs in Cal. Apr–May.

ACERACEAE: Maple Family

2.
ACER. Maple

Boxelder;
Ash-leaved Maple
Acer negundo L.
[37, 38]

Identified by opposite pinnately compound leaves with
3–5, less commonly 7–9, leaflets some of which are
coarsely and irregularly toothed; by pith less than half
the diameter of year-old twigs; by fruits having two
widely spread wings like those of other maples. Small
to medium-sized fast-growing trees to 30 m tall by 1.5
m DBH. First year twigs smooth, green; lateral buds
under petiole bases, exposed buds light-grayish hairy.
Opposite leaf scars meeting at a sharp angle around
twig or nearly so. Boxelder is quite variable over its
broad range and has been separated into several varieties and subspecies. Most botanists now recognize
none of them. Easily confused with ashes, but ashes
have entire or finely serrated leaflets and fruits with one
wing. Seedlings are easily mistaken for Poison-ivy, but
Poison-ivy has alternate leaves. Wood brittle, of little
economic value. Used as an ornamental and shade tree
but undesirable because of weedy nature, shedding of
the abundant seeds, and curling and fall of leaflets
during dry periods. See Group E for other *Acer* species.
Common. Grows in a variety of habitats but does best
in wet places such as stream banks, floodplains,
lowlands. Somewhat weedy, invading roadsides and
waste places. One of our most widely distributed trees,
extending west into Ariz, eWash, and sAlta; south into
ne, c, and sMex and Guatemala. Unusual above 600 m
in eUS, reaching 2400 m in wUS.

HIPPOCASTANACEAE: Buckeye Family

3.
AESCULUS. Buckeye

Easily recognized by opposite palmately compound leaves with 5–11 irregularly
serrated leaflets and glabrous twigs. Shrubs to large trees. Terminal buds present
except on twigs that bore flowers. Flowers showy, in large upright branched clusters. Fruits rounded, with a tough leathery covering that splits into three parts at

maturity, containing 1–3 (rarely to 6) large smooth shiny seeds. Seeds and vegetative parts poisonous when eaten. Buckeye seeds resemble chestnuts in size and
color and both have a large scar at the base. Chestnuts, however, have a pointed tip
whereas Buckeye seeds are round. Represented in the SE by 5 native species and 1
introduced from seEurope and escaped. A species occurs in Cal and a seventh in Baja
California, Mex. Five other species occur in Asia.

KEY TO AESCULUS SPECIES
1. Buds glutinous; claw of upper petals shorter than the calyx, stamens much longer
   than petals; fruits spiny or warty   6. *A. hippocastanum*
1. Buds not glutinous; claw of upper petals mostly longer than calyx; stamens
   longer or shorter than petals; fruits spiny, warty, or not   2
   2. Petals white; stamens 3–4 times the length of the petals; inflorescences
      narrow, 20–50 cm long   1. *A. parviflora*
   2. Petals yellow to red; stamens shorter than to twice the length of petals;
      inflorescences broad, 10–25 cm long   3
      3. Petals nearly equal, pale yellow to greenish-yellow; stamens about twice as
         long as petals; fruits spiny (rarely not)   2. *A. glabra*
      3. Petals unequal, the upper pair longer and narrower and with a small
         spatulate blade, yellow to red; stamens shorter than to barely longer than
         the upper petals; fruits not spiny   4
         4. Margin of lateral petals stipitate-glandular, stamens longer than the
            lateral petals   3. *A. pavia*
         4. Margin of lateral petals villous, without glands; stamens shorter than
            the lateral petals   5
            5. Pedicels minutely hairy to tomentose, without glands; stalk of
               leaflets mostly over 3 mm long   4. *A. sylvatica*
            5. Pedicels with small stalked glands; stalk of leaflets 3 mm long or less
                                   5. *A. flava*

1.
Bottlebrush Buckeye
Aesculus parviflora Walt.
[39]

Recognized by columnar flower clusters 20–50 cm
long; by white petals that are of nearly equal length
and about ⅓ as long as the stamens. Shrubs or occasionally trees to 5 m tall by 10 cm DBH. Fruits spineless, subglobose, 2.5–4 cm long, with 1–2 seeds.
Excellent as an ornamental, but not widely used. Occasional; locally abundant, usually in dense colonies due
to its frequent sprouting from roots. Rich woods of
hillsides and bottomlands. To 300 m elevation in Ala.
June–July.

2.
Ohio Buckeye
Aesculus glabra Willd.
[40]

Distinguished by broad pyramidal clusters of flowers
10–25 cm long; by pale yellow to greenish-yellow petals that are nearly equal in size and about half as long
as the stamens; by spiny fruits (rarely lacking spines).
Shrubs or trees to 44 m tall by 1.2 cm DBH. Fruits
ovoid to obovoid, 2–5 cm across, with 1–3 (rarely to
6) seeds. Wood white, easily carved and worked, but of
minor importance because of its scarcity. There are two
varieties, var. *arguta* (Buckl). Robins. having 7–11
leaflets that are 1–3 cm (rarely to 5) wide and occurring in eArk, eMo, and westward; var. *glabra* having
5–7 leaflets that are 3–6 cm wide and found almost
entirely to the east of var. *arguta*. They intergrade
where they overlap in distribution. A dwarf form
flowering when under a meter tall occurs in Douglas
Co., Ga. Twigs and leaves with an unpleasant odor
when crushed, hence sometimes called Fetid Buckeye.
Common. Rich soil of bottomlands, hillsides, and flat
uplands; usually as scattered trees mixed with other
species but sometimes in shrubby thickets along stream
banks. To about 600 m elevation. Mar–May.

3.
Red Buckeye
Aesculus pavia L.
[41]

Recognized by oblong inflorescences 10–25 cm long;
by unequal-sized petals that are usually red, often yel-
lowish-red, or yellow in central Tex, slightly shorter
than the stamens, the margin of the lateral petals with
small stalked glands; by spineless fruits. Shrubs or
trees to 12 m tall by 50 cm DBH, sometimes flowering when scarcely a meter tall. Fruits subglobose to
obovoid, 3.5–6 cm across with 1–3 (rarely to 6) seeds.
Popular as an ornamental shrub or tree. A variable
species which in the past has been divided into several
varieties and species, but more recent detailed studies
show that only the yellow-flowered plants in Tex may
possibly be considered as a separate species. Common.
Pinelands, pine-deciduous woods, wooded bluffs,
stream banks, bottomlands. To about 450 m elevation.
Feb–May.

4.
Painted Buckeye
Aesculus sylvatica Bartr.
[42]

Identified by stalk of leaflets mostly over 3 mm long; by petals yellow-green to cream and unequal, stamens shorter than the lateral petals. Trees to 15 m tall by 25 cm DBH, or more commonly shrubs. Inflorescences narrow to broadly oblong to pyramidal, 10–15 cm long; pedicels hairy but without glands; margin of the lateral petals villous and without glands; reddish flowers likely due to hybridization with *A. pavia*. Fruits globose, 2.5–4 cm across, spineless, with 1–3 (rarely to 6) seeds. Common. Most frequent as understory plants on well-drained wooded slopes, less often on stream banks and in poorly drained woods. Occurs with pines or deciduous trees. To about 400 m elevation. Apr–June. Syn: *A. georgiana* Sarg.; *A. neglecta* Lindl.

5.
Yellow Buckeye
Aesculus flava Soland.
[43]

Recognized by stalk of leaflets mostly 3 mm long or less; by petals yellow and unequal, the stamens shorter than the lateral petals. Trees to 30 m tall by 1.5 m DBH. Bark brown and scaly. Flowers in clusters 10–15 cm long, pedicels with stalked glands. Fruits 5–8 cm across. Attractive as an ornamental; wood too soft for most uses. Common. Rich mixed-deciduous woods from riverbottoms to mountain tops, occasionally in spruce-fir forests. To 1900 m elevation. Apr–June. Syn: *A. octandra* Marsh.

6.
Horse-chestnut
Aesculus hippocastanum L.
[44]

May be separated from all other naturally reproducing *Aesculus* in the SE by glutinous buds and claw of the upper petals shorter than the calyx. Trees to 25 m tall by 65 cm DBH. Leaflets 5–7. Flowers with three upper petals marked at base with yellow or red; stamens longer than petals. Fruits spiny or warty with 1–2 seeds. Rare. Planted as an ornamental, mostly as a street or park tree, escaping in scattered localities. Native of seEurope. Apr–May.

OLEACEAE: Olive Family

4.
FRAXINUS. Ash

Identified by opposite pinnately compound leaves; by 5–11 entire to prominently
serrate leaflets without lobes; by fruits with 1 wing with 2 edges (rarely with 3); by
pith of year-old twigs less than half the diameter of the twig. Species of the SE are
small to large trees. Flowers usually lacking petals, individually inconspicuous but
in prominent clusters, usually unisexual and the sexes on different trees or occasionally both sexes on the same tree or sometimes perfect. Fruits 1-seeded, dry when
mature. Since many trees have only male flowers and never produce fruits, identification to species often depends on vegetative characters except for some minute
differences in flowers. There are about 60 species of Ash in the world; 6 occur in the
SE, none being confined solely to the area; 10 species occur in wUS. Ashes are
sometimes confused with *Carya* (hickories), but the latter may be separated by their
alternate leaves. Ashes flower before leaves develop in early spring.

KEY TO FRAXINUS SPECIES
1. Youngest twigs 4-angled to narrowly 4-winged   2. *F. quadrangulata*
1. Youngest twigs rounded   2
   2. Lateral leaflets sessile, calyx absent   5. *F. nigra*
   2. Lateral leaflets with stalks 3–20 mm long or winged and appearing as short as
      1 mm; calyx present, persisting as a minute cup at base of fruits   3
      3. Leaves with dense minute rounded projections beneath, pale-colored; wing
         of fruits terminal or extending at most along a third of the body
                                    1. *F. americana*
      3. Leaves lacking minute rounded projections beneath, green to light green
         beneath but not pale-colored; wing of fruits extending on body to near the
         middle or beyond, sometimes to the base   4
         4. Stalks of basal leaflets winged nearly to base, fruit wing usually less than
            7 mm wide   3. *F. pennsylvanica*
         4. Stalks of basal leaflets wingless, fruit wing usually over 7 mm wide   5
            5. Leaf scars deeply notched; body of fruits terete or nearly so and
               winged to about midway of body   4. *F. profunda*
            5. Leaf scars not deeply notched, but may have small notch for bud;
               body of fruits flattened and winged to base   6. *F. caroliniana*

1.
White Ash
Fraxinus americana L.
[45]

Distinguished by youngest twigs being rounded; by
lateral leaflets with a stalk 3–15 mm long; by underside of leaflets pale, with dense minute rounded projections; by a small calyx at base of fruits, the wing
terminal or extending at most along a third of the
body. A variable species. Trees to 40 m tall by 2.1 m
DBH. Terminal buds obtuse, with 4–6 brownish
scales. Upper edge of leaf scars with a deep notch.
Leaflets 5–9, 6–15 cm long, glabrous or hairy beneath. Fruits 1.5–6.5 mm long, the wing 6–12 mm
wide, with a persistent calyx cup 1–2 mm long. The
wood highly valued, with many uses including veneer,
paneling, furniture, baseball bats. Fruits eaten by birds
and rodents. Common. Usually in rich deep soils,
mixed with other deciduous trees. Occurs to 1585 m
elevation in the Smoky Mts. Syn: *F. biltmoreana* Beadle. Mar–May.

2.
Blue Ash
Fraxinus quadrangulata
Michx.
[46]

Easily recognized by youngest twigs 4-angled to narrowly 4-winged. Trees to 36 m tall by 1.2 m DBH.
Terminal buds pointed, with 3 pairs of reddish-brown
scales. Leaflets 5–11, 8–13 cm long, short-stalked.
Fruits 2.5–5 cm long, the wing 5–15 mm wide and
extending to base of body. Twigs crushed and placed in
fresh water will turn the water blue, a dyestuff of the
pioneers. The wood hard, used for interior work and
floors, though sparingly because of scarcity. Fruits a
source of food for wildlife. Occasional. Usually in limestone outcrop areas, sometimes in rich deciduous
woods of valleys. Apr.

3.
Green Ash
Fraxinus pennsylvanica
Marsh.
[47]

Recognized by youngest twigs rounded; by leaflets 10–
15 cm long, rarely shorter, the lateral leaflets with a
stalk 1–9 mm long and winged nearly to base, the
underside pale green to green; by fruits with a small
calyx 0.5–1.5 mm long at base, the wing 6–8 mm
wide and extending to about the middle of the terete
body or slightly beyond. A variable species. Trees to
about 40 m tall by 1.4 m DBH, usually to about half
that size. Branches and petioles glabrous or hairy; terminal buds rounded, with 6 rusty-brown scales. Upper

edge of leaf scars nearly straight across top, or with a shallow notch. Leaflets 7–9 (rarely 5), 10–15 cm long (rarely shorter), glabrous or hairy beneath. Fruits 2.5–6 cm long. Wood with uses similar to those of *F. americana*. Common. Usually in moist to wet habitats, especially in floodplains and river terraces; mixed with other broadleaf species or occasionally in almost pure stands. To about 900 m in sAppalachians. Extends west to eCol, cWyo, sSask. Syn: *F. lanceolata* Borkh. Apr.

4.
Pumpkin Ash
Fraxinus profunda (Bush)
Bush
[48]

Identified by youngest twigs rounded, the leaf scars deeply notched; by leaflets 12–25 cm long on an un-winged stalk 5–20 mm long; by fruits with a wing 8–12 mm wide extending to about midway the body, the body over 2 mm broad, the calyx at base of fruits 2.5–6 mm long. Trees to 30 m tall and about 1.7 m DBH. Terminal buds pointed, with 3 pairs of reddish-brown scales. Wood of limited use because of scarcity. Fruits eaten by ducks and other birds. Occasional. River swamps and bottomland forests. Syn: *F. tomentosa* Michx. f. Apr–May.

5.
Black Ash
Fraxinus nigra Marsh.

Recognized by youngest twigs rounded; by sessile lateral leaflets; by calyx absent. Trees to 25 m tall by 1.4 m DBH. Terminal buds dark brown, pointed. Leaf scars nearly circular. Leaflets 7–11. Flowers usually perfect, therefore almost all trees bear fruits at some time. Fruits 2.5–4 cm long, the wing 6–15 mm wide; important food for birds, especially waterfowl. Wood tough and durable, used for woodwork, strips of split wood for baskets and woven chair bottoms. Occasional. Swamps, bogs, bottomlands, lake margins, often with spruce and fir. Extends to swNfld. Apr–May.

6.
Carolina Ash
Fraxinus caroliniana Mill.
[49]

Distinguished by youngest twigs rounded, leaf scars not deeply notched but may have small notch for bud; by leaflets pale green beneath, lateral leaflets with a slender stalk 3–20 mm long; by body of fruits flattened and winged to base, the wing 1–2 cm wide. A quite variable species. Trees to 15 m tall by 35 cm DBH, rarely larger. Trunks often several and leaning, sometimes buttressed where inundated for long periods. Buds mostly acute. Leaflets 5–9 and 8–15 cm long with a slender stalk 3–20 mm long, lower surface a paler green than upper but not whitish. Calyx on fruits about 1 mm long. Wood soft and lightweight, with minor use, mostly as pulpwood. Common. Wet habitats such as swamps, flatwoods depressions, pond margins. The southernmost of SE ashes. Syn: *F. pauciflora* Nutt. Mar–May.

VERBENACEAE: Verbena Family

5.
VITEX. Chaste-tree

Chaste-tree
Vitex agnus-castus L.
[50]

Identified by opposite palmately compound leaves with 5–7 entire leaflets; by finely hairy twigs. Shrubs or trees to 5 m tall and 18 cm DBH. Leaves glossy above, hairy beneath. Petals united, slightly unequal in size. Fruits 1-seeded, 3–4 mm long. Often planted as an ornamental, persisting and escaping at scattered localities, especially around abandoned homesites. Rare. A native of sEurope and wAsia. June–Oct.

CAPRIFOLIACEAE: Honeysuckle Family

6.
SAMBUCUS. Elder

Elderberry; American Elder
Sambucus canadensis L.
[51]

Recognized by year-old twigs with large lenticels and pith over half the diameter; by opposite once-pinnately compound leaves with 5–11 serrate leaflets, the lowest pair of leaflets sometimes deeply divided, with some forms having twice-compound leaves. Shrubs or occa-

sionally trees to 6 m tall by 27 cm DBH. Inflorescence flat-topped or nearly so, to 40 cm across. Fruits deep purple berrylike drupes, 4–6 mm long; eaten abundantly by birds and mammals and spread by their droppings. Vegetative parts eaten by a number of mammals. Those forms with twice-compound leaves have been considered by some a separate species, *S. simpsonii* Rehd. (Southern Elder), or a variety. These forms occur in the southern part of the range, especially in Fla. Common. Moist to wet places, usually in the open. From near sea level to about 1600 m elevation in Smoky Mts. There are 4 other species of tree size in wUS, some extending into Can and Mex. Mar–July.

Group D

Leaves compound and alternate.

13. Leaflets alternate, at least in upper third of leaf   16
     16. Terminal leaflet less than twice as long as wide   10. *Cladrastis*
     16. Terminal leaflet over twice as long as wide   17
        17. Leaflets lanceolate, mostly 5–13 cm long   20. *Sapindus*
        17. Leaflets elliptic, mostly 2.5–4 cm long   12. *Sophora*
12. Leaflet margin toothed, serrate, or crenate, sometimes minutely so in
     Gleditsia (use 10 × lens on underside of leaf) or with only 1–2 short
     teeth at base (*Ailanthus*)   18
     18. Leaflet margin entire except for 1–2 short gland-bearing teeth at
     base   16. *Ailanthus*
     18. Leaflet margin otherwise, any teeth at base without glands   19
        19. Pith hollow but with cross partitions (chambered)   1. *Juglans*
        19. Pith continuous, not chambered   20
           20. Leaflets with some teeth or lobes over 4 mm when
           measured perpendicular to midvein from sinus to outer
           point   21. *Koelreuteria*
           20. Leaflets with teeth under 4 mm when measured similarly
           21
              21. Apex of leaflets mostly obtuse to rounded and
              sometimes apiculate or slightly notched, stems
              usually with strong frequently branched thorns
              8. *Gleditsia*
              21. Apex of leaflets mostly acute to acuminate, twigs
              with simple prickles or unarmed   22
              22. Stems, twigs, and sometimes leaves armed with
              well-scattered to many prickles   13. *Zanthoxylum*
              22. Prickles absent   23
                23. Bruised or crushed leaves with pungent odor,
                fruits hard nuts surrounded by a 4-sectioned
                husk   2. *Carya*
                23. Bruised or crushed leaves without a pungent
                odor, fruit lacking sectioned husks   24
                24. End bud an axillary one   18. *Rhus*
                24. End bud a terminal one   3. *Sorbus*

JUGLANDACEAE: Walnut Family

1.
JUGLANS. Walnut

Identified by alternate once-pinnate leaves with 9–23 finely serrated leaflets 5–13
cm long, with a pointed tip; by pith of twigs hollow, but with prominent cross
partitions. Terminal bud present, hairy; lateral buds commonly superposed. Leaf
scars with bundle scars in 3 groups that are usually U-shaped; stipule scars absent.
Flowers unisexual, both sexes on same tree, the male in cylindrically shaped catkins

5–14 cm long, the female in a small irregularly shaped cluster at end of current year's growth. Outer wall of fruit semifleshy, indehiscent, covering a deeply corrugated pit containing, when ripe, a sweet oily kernel (the seed). There are 2 species in the SE, 4 more occur in wUS, 2 of these extending to Mex. About 12 more species occur in other parts of the world.

KEY TO JUGLANS SPECIES

1. Pith of mature twigs light brown; leaf scars of last year's leaves without a mat of hairs across the top; fruits glabrous, globose or nearly so   1. *J. nigra*
1. Pith of mature twigs dark brown; leaf scars of last year's leaves with a downy line of short matted hairs across the top; fruits glandular-hairy, ellipsoid to ovoid
2. *J. cinerea*

1.
Black Walnut
Juglans nigra L.
[52, 53]

Distinguished by twigs with a light-brown chambered pith; by leaves once-pinnately compound; by globose fruits. Trees ordinarily to 27 m tall by 1.2 m DBH, but as high as 45 m and DBH of 2.4 m. Bark dark brown to almost black, with thin anastomosing ridges. Juice of fruits containing a dark brown dye that permanently stains cloth and discolors skin. The wood one of the most highly valued of North American hardwoods, being used for such things as paneling, furniture, and gun stocks; the supply now so limited that mostly used as veneer. The dried nuts sold for food. Common. Usually in thin woods or open, but grows tall and can become a component of mature mixed-deciduous woods. Occurs to about 1175 m elevation in sAppalachian Mts. Easily escapes from plantings. Apr–May.

2.
Butternut; White Walnut
Juglans cinerea L.
[54]

Distinguished by an ashy gray trunk with broad anastomosing ridges; by pith of twigs chambered, dark brown; by a downy line of short matted hairs on upper margin of leaf scars; by leaves once-pinnately compound. Trees to 33 m tall and 90 cm DBH, but usually much smaller. Fruits ellipsoid to ovoid, pointed, usually with 2 or 4 ridges that are often obscure, with abundant sticky hairs. Wood coarse-grained, soft, used for furniture and finishing, but little available. Seeds tasty but difficult to remove from nut. Occasional. Rich soils of valleys and slopes. Shade tolerant, often growing with other deciduous trees. Apr–May.

2.
CARYA. Hickory

Recognized by twigs with continuous pith; by once-pinnately compound leaves with (3)5–19(21) leaflets that have a pungent odor when bruised or crushed; by leaflets with acute to acuminate apex, serrate margin, the teeth under 4 mm long when measured from sinus to outer point perpendicular to outline of leaflet margin, the terminal leaflet often the largest; by fruits hard nuts surrounded by a 4-sectioned husk. Small to large trees with prominent taproot, sometimes shallow-rooted in wet soils. Flowers unisexual, on same tree, appearing in spring as new leaves develop. Male flowers in long dense cylindrical catkins attached to sides and end of previous season's twigs, perianth inconspicuous, producing enormous amounts of pollen. Female flowers solitary or up to 10 in a cluster in a short spike at tip of developing twigs of current year, lacking petals, bearing a single pistil with 2 sessile stigmas. Nuts developing from the pistil, the surrounding husk from other structures associated with the flower. Nuts, which are high in unsaturated oils, an important source of food for wild mammals and birds, pastured hogs, and man (especially the widely cultivated *C. illinoensis* and much less commonly the largely wild *C. ovata*). *Carya* species provide lumber valuable for such uses as furniture, flooring, veneer, tool handles, and skis; wood high-yielding in heat, used for smoking meat. Most are highly susceptible to fire damage. There are 11 species in the US, all occurring in the SE with only 1 confined to the area; 3 extend into Mex and 5 into Can. Another species occurs only in Mex, 4 in Asia. Syn: *Hicoria*. Mar–May.

KEY TO CARYA SPECIES
1. Leaflets 5–21, often falcate, terminal leaflet smaller or about equal to adjacent ones (may be larger in No. 2); outer bud scales valvate; seams of fruit husk winged or keeled　　　2
　　2. Nuts nearly circular in cross sections, shell smooth, kernel sweet　　　3
　　　　3. Leaflets 11–21, nut shell usually thin　　　1. *C. illinoensis*
　　　　3. Leaflets 5–9, nut shell very thick　　　2. *C. myristiciformis*
　　2. Nuts flattened, shell uneven or rough, kernel very bitter　　　4
　　　　4. Leaflets 7–11, usually 7; mature terminal buds yellowish; fruit husk splitting about half way　　　3. *C. cordiformis*
　　　　4. Leaflets 7–15, usually 9–11, mature terminal buds dark reddish-brown; fruit husk splitting entire length　　　4. *C. aquatica*
1. Leaflets 3–9, not falcate, terminal leaflet larger than adjacent ones (may be nearly equal in No. 8); outer bud scales overlapping; seams of fruit husk not winged or keeled　　　5
　　5. Margin of young leaflets densely ciliate, older leaflets with persistent tufts of tiny hairs near tips of teeth; bark shaggy　　　5. *C. ovata*
　　5. Margin of young leaflets not densely ciliate, older leaflets without tufts of hairs on teeth; bark shaggy or not　　　6
　　　　6. Leaf-bearing segment of vigorous stems about 7 mm across; mature terminal buds 12–37 mm long; fruit husk 3–13 mm thick, splitting to base (tardily so in No. 7); bark shaggy or not　　　7
　　　　　　7. Bark shaggy; terminal buds 18–30 mm long, outer scales persistent until spring opening; fruit husk 7–13 mm thick　　　6. *C. laciniosa*

7. Bark not shaggy; terminal buds 12–19 mm long, outer scales soon deciduous; fruit husk 3–6 mm thick   7. *C. tomentosa*

6. Leaf-bearing segment of vigorous stems 3–6 m across; mature terminal buds 6–12 mm long; fruit husk 1–4 mm thick, splitting only partially to base; bark not shaggy   8

  8. Bud scales and undersurface of leaflets with small silvery scales; leaflets 7–9   8. *C. pallida*

  8. Bud scales and undersurface of leaflets without silvery scales, although amber-colored or yellowish scales may be present; leaflets 5–7   9

    9. Leaf-bearing portion of stems, lower surface of leaflets, and winter buds glabrous; leaflets 5   9. *C. glabra*

    9. Leaf-bearing portion of stems, lower surface of leaflets, and winter buds with abundant small amber-colored resinous scales or reddish hairs   10

      10. Leaflets bearing amber-colored resinous scales beneath and no reddish hairs; confined to Fla   10. *C. floridana*

      10. Leaflets bearing reddish hairs beneath and no amber-colored resinous scales; cLa to swInd and westward   11. *C. texana*

1.
Pecan
Carya illinoensis (Wang.)
K. Koch
[55]

Identified by the 2 outer large bud scales of terminal buds not overlapping; by 11–21 falcate leaflets that are asymmetrical at base; by husk of fruits about 1 mm thick and winged or keeled; by cross-section of nuts nearly circular in outline, the shell smooth and usually thin, kernel sweet and the halves not deeply 2-cleft. Mature trees occasionally to 35 m tall by 1 m DBH, but under cultivation recorded to 60 m by 2.5 m. Bark rough. Buds yellowish-brown, bearing small scales, lateral buds often superposed. Terminal leaflet smaller than some of the lateral ones. Clusters of male catkins on peduncles to 1 cm long. Wood light reddish-brown. Grown commercially for nuts with single trees known to produce over a half ton of nuts in a good year, but 500 pounds is average for most mature trees. Common. Most frequently in low areas, but occurs in many types of habitats. Native in area shown on map, extending in a scattered pattern to swTex and e and sMex, widely planted and escaping elsewhere. Cultivated varieties have been developed for climatic conditions for the w, e, and nUS. Escapes are rare north of Ky and Va. Apr–May. Syn: *C. pecan* (Marsh.) Engl. & Graebn.

2.
Nutmeg Hickory
Carya myristiciformis
(Michx. f.) Nutt.
[56]

Recognized by 2 large valvate outer bud scales on terminal buds; by 5–9 leaflets, the terminal one the largest or nearly so; by husk of fruits thin and winged or keeled; by nuts circular in cross-section, shell smooth and very thick, kernel sweet. Trees to about 35 m tall by 70 cm DBH. Bark usually with small thin

scales. Twigs with very small yellowish to brownish scales; buds with yellowish-brown hairs. Underside of leaflets with silvery scales. Rare. River swamps, bottomlands, and valleys; poorly drained uplands. Extends into neMex. Apr.

3.
Bitternut Hickory
Carya cordiformis (Wang.)
K. Koch
[57]

Identified by mature terminal buds yellowish with 2 large valvate outer scales; by 7–9(11) slightly falcate leaflets, the terminal one about equal in size to adjacent ones; by fruits 4-winged or ridged, often only above the middle, the husk under 1 mm thick and splitting only about half the length; by nuts a little flattened, obscurely angled, tipped with a slender persistent point, the kernel bitter. Largest tree 37 m tall by 1.2 m DBH, but usually no more than half that. Bark tight, at first smooth, finally with low narrow interlacing ridges, occasionally with a few scales. Common. Most often in low woods, less frequently in uplands. Apr.

4.
Water Hickory
Carya aquatica (Michx. f.)
Nutt.
[58]

Recognized by mature terminal buds dark reddish-brown, bud scales valvate; by leaflets 7–15, usually 9–11, terminal one smaller or about equal to adjacent ones; by fruits flattened, winged along the seams, husk splitting entire length or nearly so, kernel bitter. Trees to 35 m tall by 1 m DBH but usually considerably less. Bark of older trees light brown, splitting into long platelike red-tinged scales. Common. Swamps, along rivers. Apr–May.

5.
Shagbark Hickory
Carya ovata (Mill.)
K. Koch
[59]

Distinct among hickories in that the margin of young leaflets is densely ciliate, the older ones with persistent tufts of tiny hairs near tips of teeth. Recognized also by terminal bud scales overlapping, the outer ones blackish and persisting until bud is opening; by having 5(7) leaflets, these not falcate, the terminal one the largest. Trees to 45 m tall by 1.5 m DBH, usually around ⅔ that size. Bark with long scaly plates, the ends curving away from trunks. Fruits to about 35 mm across, husk 5–11 mm thick, freely splitting, kernel sweet. Plants with the narrower leaflets and fruits about 25 mm across are considered by some as var. *australis* (Ashe) Little and by others a separate species, *C. carolinae-septentrionalis* (Ashe) Engl. & Graebn. Common. Better drained lowlands to dry slopes, rarely in wet habitats. To over 900 m elevation in sAppalachians. Occurs in neMex. May.

6.
Shellbark Hickory
Carya laciniosa (Michx. f.)
Loud.
[60]

Recognized by very shaggy bark; by leaf-bearing segments of stem about 7 mm across; by mature terminal buds 18–30 mm long, with overlapping scales, outer scales persisting until spring opening; by leaves 35–60 cm long, leaflets 5–9, terminal leaflet the largest, the margin glabrous. Trees to 45 m tall and over 1 m DBH; bark with long thick plates with ends curving away from trunk. Fruits 45–70 mm long, husk 7–13 mm thick, nut angled, shell very thick and hard, kernel sweet. Occasional. Bottomlands and floodplains. Apr–May.

7.
Mockernut Hickory
Carya tomentosa (Poir.)
Nutt.
[61]

Identified by terminal buds 12–19 mm long, the scales hairy and overlapping, the outer ones falling soon after leaves fall; by (5)7–9 leaflets, these not falcate, the terminal one the largest, the margin lacking hairs; by fruit husk 3–6(7) mm thick and freely splitting. Trees to 35 m tall by 1.3 m DBH. Bark tight with low rounded interlaced ridges forming a netlike pattern. Leaflets with hairs on the underside curly and in tufts; fall foliage golden yellow. Fruits 35–50 mm

long, nut 4-ribbed, kernel sweet, difficult to remove because of the thick shell. Common. Dry habitats of uplands, including ridges and hillsides. To over 900 m elevation in sAppalachians. Mar—May. Syn: *C. alba* (Mill.) K. Koch.

8.
Pale or Sand Hickory
Carya pallida (Ashe)
Engl. & Graebn.
[62]

Unusual among hickories in having leaves pale beneath, although they may not be as pale as those in the photograph. Recognized also by leaf-bearing part of vigorous stems 3—6 mm across, terminal bud scales overlapping; by 7—9 leaflets, these not falcate. Trees to 35 m tall by 1 m DBH, but usually about half that. Bark with rough ridges forming a diamond pattern. Petioles bearing hairs in tufts. Fruits 13—25 mm across, the husk 2—4 mm thick and splitting to the base, nut angled and with a thin shell, kernel sweet. Occasional. Dry soils, often sandy; hillsides. To over 750 m elevation in sAppalachians. Apr—May.

9.
Pignut Hickory
Carya glabra (Mill.) Sweet
[63]

Generally can be recognized by leaves and winter buds free of rust-colored hairs and amber-colored scales; by having a pear-shaped fruit (rarely nearly globose). Positive identification also by having leaf-bearing segment of vigorous stems under 6 mm across, terminal buds 6—9 mm long, the scales overlapping; by 5, rarely 7, glabrous leaflets. Trees to 40 m tall by 1.8 m DBH, but usually about half that. Bark of young trees smoothish, later with tight anastomosing ridges forming a diamond pattern. Fruit husk 1.5—2.5 mm thick, splitting about half way, uncommonly remaining closed or promptly splitting to base; shell of nut about 2 mm thick, kernel usually bitter. Plants with fruit husk promptly splitting to base have been separated as var. *odorata* (Marsh.) Little and as a species, *C. ovalis* (Wang.) Sarg. There are now generally considered to be too many intermediates for recognition at either level. Common. Usually in drier habitats, from dunes to hillsides and rocky areas. Shade tolerant. Apr–May. Syn: *C. leiodermis* Sarg.

10.
Scrub Hickory
Carya floridana Sarg.

Having characteristics of *C. glabra* except leaflets are 3–5(7) with amber-colored resinous glands on under-surface, and fruits less often pear-shaped. Intergrades with *C. glabra*. Rare. Scrub areas of sand ridges, Marion Co., Fla, and southward. Apr.

11.
Black Hickory
Carya texana Buckl.
[64]

The only hickory with tufted rust-colored hairs on twigs, buds, petioles, and lower surface of leaflets, usually almost all the hairs lost late in growing season. Positive recognition also by having leaf-bearing segment of twig under 6 mm across, bud scales overlapping; by 5–7 leaflets. Trees to 30 m tall by 0.9 m DBH, but usually less than half as large. Bark rough, with irregular blocky ridges. Fruits globose to broadly obovoid, husk about 3 mm thick, kernel sweet. Occasional. Dry sandy areas, rocky slopes and ridges. Apr.

ROSACEAE: Rose Family

3.
SORBUS. Mountain-ash

American Mountain-ash
Sorbus americana Marsh.
[65]

Recognized by end bud a terminal one, glabrous twigs with a large reddish-brown, continuous pith and no prickles; by once-pinnately compound leaves with 9–17 sessile serrate leaflets, the teeth acuminate; by fruits small orange-red pomes. Shrubs or trees to 18 m tall by about 76 cm DBH. Terminal buds large, with a curved pointed tip. Flowers in compound corymbs, the clusters at a distance resembling those of *Sambucus*. Fruits bitter, persistent into winter, valuable as food for birds and rodents. Fruits and leaves colorful in fall; used as an ornamental, but not adapted to warmer habitats. Common. In open or in woods. At high elevations, rarely below 950 m in sAppalachians; to sea level farther north. From swOnt, to Lab and Nfld, but absent from southernmost Ont. A similar species occurs from Minn to Conn and Lab; 2 species occur in wUS and Can, extending to sAlas. June–July. Syn: *Pyrus americana* (Marsh.) DC.

FABACEAE: Legume Family

4.
ALBIZIA. Albizia

Mimosa-tree; Silk-tree
Albizia julibrissin Durz.
[66]

Identified by leaves twice-pinnately compound, 15–35 mm long; by leaflets many, 3–5 mm wide, strikingly one-sided, the main vein near one margin. Trees to about 15 m tall, often with an umbrella-shaped crown. Twigs zigzag, with odor of green peas when broken; terminal buds absent; axillary buds superposed, partly embedded in twig. Swollen areas located at base of petioles, leaf segments, and leaflets sensitive to light; as light diminishes the leaflets fold against each other and the leaves and their segments droop. Flowers in nearly globose inflorescences about 4–6 cm across. Stamens are so prominent that close inspection is usually necessary to see the 5-parted calyx and deeply 5-lobed corolla. Fruits thin flat indehiscent pods. Seeds hard, in a single row, often remaining viable on or in the soil for several years. Injured or killed by severe winters; subject to a "wilt" that is fatal. Often used as an ornamental. Occasional. Escaping to open areas and thin woods essentially throughout the SE. Usually absent above 1000 m elevation. Native of Asia. Apr.–Aug.

5.
ACACIA. Acacia

Sweet Acacia
Acacia farnesiana (L.)
Willd.

Having flower clusters similar to those of *Albizia julibrissin,* but yellow and only about 1 cm across. Leaves of similar structure, but mostly under 15 cm long and the leaflets under 2 mm wide. Shrubs or trees with spiny branches. Rare. Sandy soils in open or thin woods; sLa, Fla, islands of Ga. Apr–Nov.

6.
PROSOPIS. Mesquite

Honey Mesquite
Prosopis glandulosus Torr.
[67]

Recognized by glabrous twice-pinnately compound leaves with 2 or rarely 4 pinnate divisions, each division with 12–40 leaflets, petioles long and slender; by leaflets sessile, entire, acute at apex, 2 mm or more wide, and 7–15 × as long. Shrubs or trees to 20 m tall by 1.2 m DBH, usually much smaller. Branches with spines, usually in pairs. Flowers in axillary spikes. Fruits essentially straight pods 7–20 cm long and about as thick as broad. Flowers a source of nectar for honey; fruits eaten by various animals. Rare. In nwLa, probably introduced long ago along cattle trails. Common to the west, Tex to sCal and sMex. Feb–Apr mostly, but sporadically at other times.

7.
PARKINSONIA. Parkinsonia

Jerusalem-thorn
Parkinsonia aculeata L.

Having leaves of similar structure, 2 or 4 pinnate divisions, but the petioles quite short and the leaflets under 2 mm wide. Shrubs or small trees with spiny branches. Flowers about 1 cm across, sparsely clustered in axillary racemes; petals bright yellow, tinged with orange, and slightly unequal. Fruits crooked pods 5–10 cm long, constricted between seeds. Rare. Planted and escaping along coast from Tex to Ga and in Fla. Mar–frost.

8.
GLEDITSIA. Honey-locust

Honey-locust
Gleditsia triacanthos L.
[68]

Recognized by leaves once- or twice-pinnately compound; by apex of leaflets mostly obtuse, rounded, or slightly notched, the blade margin shallowly crenate, often nearly entire. Trees to 45 m tall and 1.6 m DBH, usually much smaller. Stems, including trunk, with strong thorns that are frequently branched; thorns uncommonly absent. Terminal buds absent; axillary buds minute, seemingly absent or indistinct, often several superposed at each node, being submerged in leaf scar or hidden by it. Fruits flat, 15–55 cm long, 2.0–3.5 cm broad, sometimes remaining on trees in winter;

seeds many, arranged in one row. Wood hard, tough, durable, and thus good for fenceposts. A thornless form is widely used as an ornamental, frequently along streets and around public buildings. Fruits eaten by animals, the seeds scattered widely. Common. Bottomlands and uplands, in open or in woods. Planted and escaping beyond natural range, as far as NS. Uncommon above 600 m elevation. Apr–June.

Water-locust
Gleditsia aquatica Marsh.

Quite similar to *G. triacanthos* and can scarcely be separated except by fruits which are well under 15 cm long and have only 1, or rarely to 3, seeds. Trees to 30 m tall and 73 cm DBH. Wood also durable. Common. Wet habitats. Apr–June.

9.
GYMNOCLADUS. Coffeetree

Kentucky Coffeetree
Gymnocladus dioicus (L.)
K. Koch
[69]

Recognized by large twice-pinnately compound leaves; by many opposite entire leaflets, these over 2 cm wide, the tips acute. Trees to about 30 m tall by 1.5 m DBH. Bark thick, with narrow scaly ridges. Twigs stout, pith of larger ones salmon-colored. Terminal buds absent; axillary buds small, downy, 2–3 superposed at each node, notably separated from leaf base. Leaves to 1 m long, developing late in spring and dropping early. Male and female flowers on separate trees; petals 4–5, greenish-white, spreading, 8–10 mm long. Fruits pods 8–25 cm long and 3–5 cm broad, hard, leathery, falling without opening in winter or sometimes later. Seeds in one row, poisonous when eaten. Sparingly planted as an ornamental and sometimes escaping. Wood used locally for fenceposts, cabinets, and general use. Occasional. Damp rich soils of lowlands and rich soils of ravines. Another species grows in China. Apr–June.

10.
CLADRASTIS. Yellowwood

Yellowwood
Cladrastis kentukea
(Dum.-Cours). Rudd
[70]

Identified by once-pinnately compound leaves; by leaf-lets alternate, 3–5 cm wide, the margin entire. Trees to 18 m tall by 80 cm DBH with thin smooth gray bark. Terminal buds absent; axillary buds small, with-out scales, woolly, superposed but usually appearing as one, usually hidden by petiole. Leaflets 5–11, turning yellow in autumn. Flowering abundantly some years; flowers in terminal panicles to 35 cm long, the corolla white and shaped like a pea flower. Fruits 6–10 cm long, flat, narrow, usually 4–6 seeded, seeds in a sin-gle row. Prized as an ornamental. The wood brown, durable, valuable partly for its scarcity for such uses as gunstocks, trays, bowls, and paneling. Rare. Rich rocky coves of mountains, limestone cliffs, rich hard-wood forests. To about 1080 m elevation. Only species of genus. Apr–June. Syn: *C. lutea* (Michx. f.) K. Koch.

11.
ROBINIA. Locust

Shrubs or trees, with 2 spines (stipules) at some or all nodes, these sometimes absent, especially on slow-growing trees. Terminal buds absent; axillary buds downy, without scales, superposed, hidden by petiole bases. Leaves once-pinnately com-pound. Leaflets 7–19, entire, opposite, over 6 mm wide, the apex obtuse to rounded or notched. Flowers shaped like pea flowers, in racemes in axils of leaves of current year's growth. Fruits flat short-stalked pods with a few to about 10 seeds, rarely more, in one row. Wood hard and durable, frequently used for fenceposts, less commonly for fuel and handicrafts. Sometimes grown as an ornamental, but can become obnoxious due to spreading by root sprouts. Seeds and vegetative parts are poisonous when eaten. There are 3 species in the SE that are trees; other species are shrubs and the number is in dispute. One tree species occurs in swUS and adj. Mex. Apr–June.

KEY TO ROBINIA SPECIES

1. Flowers 15–25 mm long, petals white, calyx lobes shorter than tube; twigs of current year glabrous   1. *R. pseudoacacia*
1. Flowers 20–25 mm long, petals pink to rose-colored or purplish, calyx lobes equalling or longer than tube; twigs of current year with sticky sessile glands and/or glandless hairs to 5 mm long   2
   2. Twigs with sticky sessile or raised glands, leaflets usually 13–21   2. *R. viscosa*
   2. Twigs without sticky glands, frequently with glandless hairs to 5 mm long, leaflets usually 9–13   3. *R. hispida*

1.
Black Locust
Robinia pseudoacacia L.
[71]

Different from all other *Robinia* species by having glabrous twigs and white petals. Trees to 29 m tall by 2.2 m DBH, rarely a shrub. Trunk straight or crooked; bark to 4 cm thick, with long rough forking ridges. Leaflets 7–19, entire, 1–2 cm wide. Flowers fragrant, 15–25 mm long, in drooping clusters, calyx lobes shorter than tube. Fruits 5–10 cm long, 11–13 mm broad. Wood durable, abundantly used for fence-posts and widely planted for this purpose. Also used as an ornamental, for shelter belts in drier areas, and for erosion control. Most likely tree-sized *Robinia* to be encountered. Common. In woods, but generally absent from rich dense woods; also in fields and other open moist to dry habitats. To about 1000 m elevation in sAppalachians. Well established beyond its natural range. Feb–June.

2.
Clammy Locust
Robinia viscosa Vent. ex Vaug.

Different from all other *Robinia* trees in having sticky glands on twigs, petioles, flower stalks, and fruits. Shrubs or small trees to 10 m tall by 25 cm DBH. Leaflets usually 13–21. Petals light to dark pink, rarely white. Rare. Thin woods, open places. May–June.

3.
Pink Locust
Robinia hispida L.
[72]

Identified by absence of sticky glands; by petals pink to rose-colored or purplish; by calyx lobes equaling or longer than the tube. Shrubs or uncommonly trees, to 5 m tall by 20 cm DBH. Uncommonly planted as an ornamental. Rare. Thin woods, especially on slopes and ridges. To about 1600 m elevation in sAppalachians. Apr–July. Syn: *R. boyntonii* Ashe; *R. fertilis* Ashe; *R. kelseyi* Hort. ex. Crowell; *R. nana* Ell.; *R. rosea* Ell.

12.
SOPHORA. Sophora

Texas Sophora
Sophora affinis T. & G.
[73]

Recognized by absence of spines; by leaves once-pinnate; by leaflets 13–19, 2.5–4 cm long, toward tip some alternate, terminal one about 2 × as long as wide. Shrubs or trees to 10 m tall by 48 cm DBH. Flowers about 12 cm long, shaped like pea flowers, in drooping lateral racemes; petals white tinged with pink. Fruits 2–8 cm long, resembling a string of beads, each bead about 1 cm across. The red seeds are sometimes made into beads, but plants of little economic value. Rare. Along streams, in limestone uplands. Apr–June.

RUTACEAE: Citrus Family

13.
ZANTHOXYLUM. Prickly-ash

Shrubs or trees recognized by twigs, and often leaves, armed with prickles, these sometimes quite scattered; by once-pinnately compound leaves with 7–22 leaflets, these over 1 cm wide, with tiny transparent dots and margin shallowly crenate to rarely nearly entire. Flowers small, of one sex, the male and female on separate plants, occasionally on same plant; petals 4–6; pistils 1–3. Fruits splitting along one side, with one shiny black seed. A large genus of many species, mostly tropical. In N. Amer there are 6 species of tree size, 2 occurring in the SE, 4 in the US outside the range of this book. Syn: *Xanthoxylum*.

KEY TO ZANTHOXYLUM SPECIES
1. Leaflets membranous, with very small hairs beneath, these sometimes scarce late in season; flower clusters axillary   1. *Z. americanum*
1. Leaflets coriaceous, glabrous beneath; flower clusters terminal

                                                                2. *Z. clava-herculis*

1.
Common Prickly-ash
Zanthoxylum americanum
Mill.
[74]

Recognized by leaflets membranous; by axillary flower and fruit clusters. Shrubs or less commonly trees to 10 m tall by 15 cm DBH, often sprouting from roots. Twigs with prickles, these sometimes sparse. Buds hairy. Leaves bitter-aromatic, the axis usually prickly;

leaflets 5–11, opposite, hairy beneath when young, sometimes nearly glabrous late in season, the margin shallowly crenate, sometimes minutely so. Fruit pods stalked. Wood of no commercial value; an oil that occurs throughout the plant, especially the bark, has medicinal qualities, but extensive folk use can eradicate this already rare species. Rare, but common to the north. Moist to dry places. In thin woods or open. Apr–May.

2.
Hercules-club
Zanthoxylum
clava-herculis L.
[75]

Recognized by having glabrous leathery leaves; by flowers and fruits in terminal clusters. Shrubs or trees to 12 m tall by 70 cm DBH. Trunk of younger trees grayish, with pyramidal corky knobs topped by a sharp prickle. Prickles present on leaf axis and usually on twigs. Buds glabrous. Leaves and twigs aromatic when crushed. Leaflets 7–19, commonly 7–9, the margin crenate, sometimes minutely so. Female flowers with 1–3 pistils. Fruit pods sessile or nearly so. Wood of no commercial value; an oil that occurs throughout the plant, especially the bark, has medicinal qualities, but folk use can threaten the species with extinction. Occasional. Dunes, maritime woods, lowland woods, hammocks, fencerows. Mar–May.

14.
PTELEA. Hoptree

Common Hoptree
Ptelea trifoliata L.
[76]

Identified by absence of thorns and by having 3 leaflets. Musky-aromatic shrubs or trees to about 9 m tall by 24 cm DBH. Lenticels large, brown, becoming horizontally elongaed. Terminal buds absent; axillary buds small, silvery-silky, partly concealed. Leaves alternate or a few sometimes opposite or nearly so. Leaflets varying considerably in size and shape, with tiny translucent dots, the terminal leaflet sessile and thus different from Poison-ivy and Poison-oak, in which the terminal leaflet is stalked. Flowers small, greenish-white, in terminal clusters, foul-smelling. Fruits flat, broadly winged, with 2–3 seeds. A number of subspecies and varieties have been named. Of little value. Occasional. Rocky woods, stream terraces, and better-drained lowlands. Extends westward into sCol and Ariz, south into sMex. Another species occurs in Cal. Mar–May.

15.
PONCIRUS. Trifoliate Orange

Trifoliate Orange
Poncirus trifoliata (L.) Raf.
[77]

Recognized by having branches armed with thorns, petioles winged, and 3 leaflets. Shrubs or trees to 7 m tall. Flowers in axillary clusters; petals white, 15–20 mm long. Fruits about 3–5 cm across, similar to a small orange except finely hairy, strongly aromatic, very sour, and with little pulp. Impenetrable as a hedge. The only other species with 3 leaflets in this group is *Ptelea trifoliata*. Rare. Planted, persisting, and spreading locally; fencerows. Mar–Aug.

SIMAROUBACEAE: Quassia Family

16.
AILANTHUS. Ailanthus

Tree-of-heaven;
Ailanthus
Ailanthus altissima (Mill.)
Swingle
[78]

Recognized by once-pinnately compound leaves; by having 11–41 leaflets, these 2.5–5 cm wide, with entire margin except for 1–2(3) short gland-bearing teeth at base on each side, terminal leaflet sometimes absent. Vigorous trees to 25 m tall by 1.8 m DBH. Twigs robust, velvety hairy, terminal buds absent; axillary buds partly concealed by petiole base. Pith large, brownish or sometimes pale. Flowers small, yellowish-green, in large terminal clusters; male and female flowers on separate trees or on the same tree, or flowers perfect. Fruits 1-seeded, 1–6 from a single flower. Widely planted, but undesirable in several respects: male flowers and bruised foliage have an unpleasant odor, persistent sprouting from roots makes control difficult, and winged fruits result in widespread dissemination. Contact with leaves causes dermatitis in some individuals. Occasional; uncommon in CP from Tex to NC. In dry to wet places and in woods or open; vacant lots, crevices of sidewalks and masonry. From nFla to Tex, seKan, sOnt, and Me. Native of China. Ten species occur in Asia and in nAustralia. May–July.

MELIACEAE: Mahogany Family

17.
MELIA. Chinaberry

Chinaberry
Melia azedarach L.
[79]

Identified by leaves twice-pinnately compound, or occasionally some thrice-compound, without prickles, the petiole of the largest leaves over 6 cm long; by leaflets 2–8 cm long by 9–19 mm wide, the margin coarsely serrate to irregularly cut or lobed, the apex acute to acuminate. Trees to 15 m tall by 60 cm DBH. Twigs stout, pith large and white, terminal buds absent. Flowers fragrant, in axillary clusters, petals lavender or rarely white. Mature fruits yellow globose drupes, 10–15 mm across, the stone with 5–8 prominent ridges and containing 1–6 seeds. Fruits poisonous when eaten; plants contain azadirachtim, an insecticide. Widely planted as an ornamental, otherwise of little value. Occasional. Spreading extensively by seeds. In woods or open, fencerows. Native of Asia. Escaped in Cal and probably other swUS states. Mar–May.

ANACARDIACEAE: Cashew Family

18.
RHUS. Sumac

Recognized by end bud an axillary one, pith of twigs of current year continuous and about half the diameter of the twig or larger; by leaves once-pinnately compound; by leaflets toothed, basal teeth glandless, if entire the leaf rachis winged. Unarmed shrubs or trees. Bark with horizontal lenticels, twigs glabrous or hairy, sap often milky. Axillary buds small, without scales, hairy, often partly hidden by petiole base. Leaflets opposite, with acute to abruptly acuminate apex, margin entire to crenate or serrate. Male and female flowers usually on separate plants. Fruits small drupes, 1-seeded. A genus of about 120 species, mainly of Africa. There are 11 species of tree size in the US; 3 occur in the SE, but extend elsewhere. All three have flowers and fruits in terminal clusters, unlike Poison Sumac, *Toxicodendron vernix*, in which clusters are axillary.

KEY TO RHUS SPECIES

1. Twigs of current year glabrous, leaf rachis not winged   1. *R. glabra*
1. Twigs of current year with dense fine or velvety hairs, leaf rachis winged or not winged   2
     2. Twigs with dense fine hairs, leaf rachis with wings, these sometimes inconspicuous   2. *R. copallina*
     2. Twigs with dense velvety hairs, leaf rachis without wings   3. *R. typhina*

1.
Smooth Sumac
Rhus glabra L.
[80]

Recognized by current year's twigs stout, glabrous, and somewhat 3-sided; by leaflets shiny green above, glaucous beneath, the margin serrate, tip of serrations mostly acute. Shrubs or trees to 12 m tall by 18 cm DBH, sprouting abundantly from roots and often forming thickets. Twigs usually glaucous, sap milky. Leaf base almost encircling the bud; leaves red in the fall. Flowers about 3 mm across. Fruits 3–5 mm long, dark red, covered with short hairs; some fruits often remaining through the winter, but usually eaten by wildlife. Occasionally planted. Common. In a variety of open habitats, wet to dry, especially common in old fields, fencerows, wood margins. To about 1350 m in SE. Occurs in all 48 conterminous states, but also grows in nMex and sCan west to BC. May–July.

2.
Shiny Sumac;
Winged Sumac
Rhus copallina L.
[81]

Distinguished by dense fine hairs on twigs of current year; by leaf rachis winged, the wings sometimes inconspicuous. Shrubs or trees to 16 m tall by 25 cm DBH, abundantly sprouting from roots, often forming thickets. Juice not milky. Buds about half encircled by leaf scars; lenticels prominently raised. Underside of leaflets with many fine hairs, the margin entire or occasionally crenate to finely serrate. Leaves turning red in autumn. Fruits 3–5 mm long, reddish-brown, covered with fine hairs, persisting into winter. Sometimes planted as an ornamental and to attract birds. Fruits eaten by birds, leaves and fruits by deer, the bark and twigs by rabbits, especially in winter. Common. Usually in well-drained habitats. Woods; especially in open places such as old fields, roadsides, and fencerows. To about 1400 m elevation. June–Sept.

3.
Staghorn Sumac
Rhus typhina L.
[82]

Identified by twigs densely and prominently hairy, remaining so for several years. Low shrubs, as in photograph, or trees to 15 m tall by 32 cm DBH, usually considerably smaller, sprouting abundantly from roots and forming thickets. Juice milky, turning black when exposed to air. Leaves a brilliant orange to red in autumn. Leaflet margin prominently serrate. Fruits 2–5 mm across, covered with red hairs. Grown as an ornamental. Eaten by animals, as in *R. copallina*. Common. Dry to moist habitats; thin woods, roadsides, old fields, other open areas. To over 1500 m elevation in Appalachian Mts. Extends into nNS. May–July.

19.
TOXICODENDRON. Poison Sumac

Poison Sumac
Toxicodendron vernix (L.)
Kuntze
[83]

Recognized by once-pinnately compound leaves without a winged rachis; by leaflets 7–13, opposite, over 1 cm wide, margin entire, apex acute to acuminate. Shrubs or trees to 7 m tall by 35 cm DBH. Bark light to dark gray. Sap watery, darkening upon exposure to air. Terminal buds present, with 2 purplish scales. Axis of leaf and leaflet stalks reddish. Leaves turning orange to scarlet in autumn. Flowers usually unisexual, the male and female on the same plant. Common name derives from poisonous nature of plant upon contact, resulting in painful skin rashes for most individuals. This plant can be further distinguished from the non-poisonous *Rhus* species it so closely resembles by its axillary flower and fruit clusters (as opposed to terminal clusters) and by fruits that are glabrous, light-colored to almost white, and about 6 mm across. Fruits are eaten by wildlife. Common. Moist to wet places in woods or open. Uncommon in the mountains. Apr–June. Syn: *Rhus vernix* L.

SAPINDACEAE: Soapberry Family

20.
SAPINDUS. Soapberry

Western Soapberry
Sapindus drummondii
Hook. & Arn.
[84]

Recognized by once-pinnately compound leaves; by leaflets mostly 5–13 cm long, over twice as long as wide, one-sided, falcate, lanceolate, margin entire. Large shrubs or trees to 20 m tall by 66 cm DBH. Bark gray to tan and wood yellow. Terminal buds absent; axillary buds usually superposed, upper bud the larger. Leaflets all alternate, or some alternate and some opposite on the same leaf, or all opposite on an occasional leaf. Foliage an attractive yellow in autumn. Flowers and fruits in conspicuous terminal panicles. Fruits berries about 13 mm across, globose, with amber-colored to yellowish translucent flesh surrounding the dark globose seed. Leaves and part of the fruits poisonous when eaten. Occasional. Upland limestone sites, fields, edge of woods, fencerows, along streams. Extending into seCol, cNMex, ceAriz, and into adj. nMex. A genus of about 40 species, mostly tropical, with 2 occurring in the US. Mar–July.

Florida Soapberry
Sapindus marginatus L.

Similar to *S. drummondii*, but fruits about 20 mm across and lopsided, having on one side a nearly globose part containing the seed and on the other side 2 dried-up aborted lobes. The widely separated distributions aid in recognition. Rare. Hammocks of Fla and woods of coastal islands in Ga. Apr–May.

21.
KOELREUTERIA. Koelreuteria

Goldenrain-tree
Koelreuteria paniculata
Laxm.
[85]

Recognized by twice-pinnately compound leaves, the smallest ones sometimes once-pinnate; by leaflets 2–4.5 cm wide, with acute to short-acuminate apex, irregularly serrated to lobed margin, terminal leaflet sessile or nearly so. Spineless round-topped trees to 15 m tall. Petals a bright yellow. Fruits bladdery capsules 2.5–5 cm long, splitting open into 3 parts. Commonly planted as an ornamental, flowering when few other trees are in bloom and having attractive fruits as well. Rare escape near plantings. A genus of 4 species of eAsia. May–June.

ARALIACEAE: Ginseng Family

22.
ARALIA. Aralia

**Devil's Walkingstick;
Hercules-club**
Aralia spinosa L.
[86]

Identified by large twice- or thrice-pinnately compound leaves with prickles on the petiole and main axis; by ovate to broadly elliptic leaflets 3–10 cm long. Little-branched shrubs or trees, often unbranched, to 11 m tall by 70 cm DBH, often forming dense colonies. Trunk with strong prickles, unless quite large. Pith very large, white. Leaves to 1 m long. Flowers and fruits in large showy terminal clusters, the ultimate divisions being small umbels. Flowers about 3 mm across with 5 white petals, often of one sex. Fruits berrylike drupes about 6 mm across with 3–5 seedlike stones. Planted as an ornamental, but difficult to contain because of sprouting from roots. Insects, especially bees, abundant around flowers. Fruits, which ripen in autumn, prized by birds and other animals. Common. Dry to moist places, more often the latter, in rich or thin woods of various types. From near sea level to over 1500 m elevation in sAppalachians. A genus of about 30 species, mostly herbs or shrubs, *A. spinosa* the only tree represented in the US. May–Aug.

Group E

Leaves simple and opposite or whorled.

12. Tip of leaves long acuminate, surfaces glabrous except on upper midrib; fruits drupes   6. *Forestiera acuminata*

12. Tip of leaves obtuse to short acuminate, surfaces glabrous or hairy; fruits capsules or drupes   13

    13. Bud scales sharply keeled and with an abrupt stiff sharp point; largest mature leaves over 35 mm wide   7. *Chionanthus*

    13. Bud scales not sharply keeled and lacking stiff sharp points, if pointed not abruptly so; largest mature leaves 8–80 mm wide   14

        14. Bundle scars in leaf scar 3   14. *Viburnum*

        14. Bundle scar in leaf scar 1   15

            15. Longest petioles about 5 mm long or less; petals white to cream-colored, united; fruits berrylike drupes with 1–4 seedlike stones; pith of first year twigs greenish   8. *Ligustrum*

            15. Longest petioles about 10 mm long or more; petals maroon to dark purple, separate; fruits smooth unequally 4-lobed capsules splitting open when mature, usually exposing the 4 red seeds; pith of first year twigs whitish   2. *Euonymous*

MORACEAE: Mulberry Family

1.

BROUSSONETIA. Paper-mulberry

Paper-mulberry
Broussonetia papyrifera (L.)
Vent.
[87]

This species is variable in its leaf arrangements and it is therefore important to examine a number of twigs for identification. On a single tree there may occur some vigorous twigs with opposite and/or whorled leaves only, other vigorous twigs with alternate leaves only, and still other vigorous twigs with opposite and/or whorled and alternate leaves all combined. A single tree may have only opposite and/or whorled leaves on all its twigs, or a single tree may have all leaves alternate. A Paper-mulberry having all leaves opposite and/or whorled, with perhaps a few alternate ones present, may be identified by leaves with 3 main basal veins; by twigs of current year densely hairy and with a soft continuous pith except for a firm greenish area at the nodes. On specimens having alternate leaves, the species may be distinguished by the absence of thorns;

by leaves rough hairy above and velvety hairy beneath, with 3 prominent veins at the base, more than 2-ranked; and by vegetative parts other than spicy-aromatic. Large shrubs or trees to 15 m tall. Twigs with milky sap; terminal buds absent. Leaves 2–3-lobed or unlobed. Flowers unisexual, the sexes on separate trees; male flowers in catkins similar to those of *Morus;* female flowers in hanging globe-shaped structures. Mature fruits partly reddish, as seen in photograph. Planted and sprouting profusely from roots, often forming large colonies. Most ornamental plantings are of male trees, occasionally of female trees, but rarely the two sufficiently close to produce fruit. Common. Waste places, fencerows, roadsides, vacant lots. A genus of perhaps 2 species of eAsia. Mar–Apr.

CELASTRACEAE: Bittersweet Family

2.

EUONYMUS. Burningbush

Burningbush
Euonymus atropurpureus
Jacq.
[88]

Identified by current year's twigs green and 4-angled or 4-winged; by bud scales not keeled; by leaves pinnately veined and deciduous. Shrubs or less commonly trees to 8 m tall by 18 cm DBH. Terminal buds ovoid, glabrous; leaf scars with 1 bundle scar. Leaf apex short-acuminate, petioles to about 10 mm long, stipules minute. Flowers about 10 mm across with 4 dark red to purple petals. Fruits smooth, deeply 4-lobed, splitting open and exposing seeds covered by a red aril. Occasionally planted for the attractive red leaves and seeds in autumn. Seeds eaten by birds and vegetation by deer, but fruits and vegetative parts may be poisonous when eaten. Occasional. Wet woods, rich wooded hillsides, stream banks. Rare over 600 m elevation. A genus of about 175 species, mostly in Asia; 4 species native to N. Amer, only this species and one in Wash to Cal attaining tree size. May–July.

ACERACEAE: Maple Family

3.
ACER. Maple

One maple species, *Acer negundo,* has the characteristics of Group C and is described
there. Maples of Group E are identified by current year's twigs glabrous and with a
continuous pith; by palmately veined leaves that are characteristically palmately
lobed, rarely unlobed and then under 10 cm long and with serrate margin. Leaf scars
opposite and meeting or connected by a transverse ridge. Flowers unisexual or, if
perfect, usually functioning unisexually. Fruits always two-winged samaras, each
wing 1-seeded; eaten by birds and rodents. Very important source of lumber for
many uses such as flooring, furniture, cabinets, gunstocks, and veneer. Another
important product is maple syrup made from the sap of *A. saccharum* primarily, to
some extent from *A. nigrum,* and in the past from *A. floridanum,* but we know of no
current production from this last one. A genus of over a hundred shrubs and mostly
trees of Asia, Europe, nAfrica, and N. Amer; 9 species (2 of these introductions)
occur in eUS, 4 in wUS of which 3 extend into swCan. *Populus alba* sometimes has
leaf blades lobed like those of some maples, but may be separated by alternate
leaves.

KEY TO ALL ACER SPECIES
1. Leaves compound (described under Group C)   *A. negundo*
1. Leaves simple   2
  2. Leaves unlobed (rare CP form)   1. *A. rubrum*
  2. Leaves lobed   3
    3. Middle lobe of most or all leaves wider near middle than at base, the
      margin with several teeth   3. *A. saccharinum*
    3. Middle lobe of most or all leaves widest at the base, if not, the lobe with
      only a few teeth or none   4
      4. Buds distinctly stalked, with 2 opposing scales   5
        5. Leaves coarsely toothed, about 2−3 teeth per cm on the larger leaves;
          clusters of flowers and fruits upright   4. *A. spicatum*
        5. Leaves finely toothed, usually 5−10 teeth per cm on the larger leaves;
          clusters of flowers and fruits hanging   5. *A. pensylvanicum*
      4. Buds sessile, with more than two scales   6
        6. Leaf margin irregularly serrated   7
          7. Each flower and fruit on a separate stalk arising essentially from a
            common point and forming a close rounded cluster. Fruits 15−26
            mm long, maturing in spring   1. *A. rubrum*
          7. Each flower and fruit on a branch of a long-stalked and elongated
            cluster. Fruits 30−50 mm long, maturing in summer. Rare escape
            from plantings   2. *A. pseudoplatanus*
        6. Leaf margin lacking serrations, but usually with a few large teeth   8
          8. Petioles exuding milky juice when broken from twig (Rare escape
            from plantings)   6. *A. platanoides*
          8. Petioles lacking milky juice   9

9. Leaves hairy and yellowish-green beneath, ends and sometimes sides of lobes drooping   10

     10. Trees with 2 to several trunks, sometimes one, the trunk(s) at least slightly crooked; bark light-colored, almost white   7. *A. leucoderme*

     10. Trees with a single straight trunk, bark dark brown or dark gray to nearly black   8. *A. nigrum*

9. Leaves glabrous or hairy, light green or pale to heavily glaucous beneath, flat or with tips and sides of lobes drooping   11

     11. Leaves glabrous beneath or rarely conspicuously hairy or occasionally hairy only in axils of main veins, tips of lobes sharply acute to acuminate. Fruits 25–30 mm long   9. *A. saccharum*

     11. Leaves hairy beneath, tips of lobes usually acute to rounded. Fruits 20–25 mm long   10. *A. barbatum*

1.
Red Maple
Acer rubrum L.
[89, 90]

Identified by sessile buds with at least 2 pairs of exposed scales; by leaf margin irregularly serrated; by blades with 3–5 lobes, the terminal lobe widest at the base, or unlobed in a rare CP form. Trees to 37 m tall by 1.6 m DBH. Terminal buds blunt. Leaves glabrous to hairy, underside light green to heavily glaucous; in autumn yellow, orange, red to purplish-red. The photograph with red leaves is a 3-lobed form of the CP with some leaves scarcely lobed. Fruits maturing in the spring, yellow to red or finally reddish-brown. Often used as an ornamental, flowers and fruits being showy in spring and leaves in autumn. Wood heavy, but not as hard or highly prized as that of *A. saccharum;* used for such things as furniture, cabinets, veneer, flooring. Easily confused with *A. saccharinum,* which is separated by having the terminal lobe constricted toward its base. Common. In wet to dry places, dense woods or in open. To over 1800 m elevation in sAppalachians down to sea level. Extends to sNfld. Jan–May. Syn: *A. drummondii* Hook. & Arn. ex Nutt.

2.
Sycamore Maple
Acer pseudoplatanus L.

Also having sessile buds, serrated leaf margin, and the middle lobe widest at base. It is separated by flowers and fruits on branches of long-stalked clusters, whereas those of *A. rubrum* are on separate stalks arising essentially from the same place and forming a close cluster. Rare escape from plantings. May.

3.
Silver Maple
Acer saccharinum L.
[91]

Easily recognized by middle lobe of most or all leaves narrower at base than near the middle of the lobe, the margin irregularly toothed. Trees of rapid growth, to 38 m tall by 2 m DBH, but usually much smaller. Broken twigs with a notable rank odor; terminal buds blunt. Leaves silvery-white beneath, conspicuous in a breeze. Flowering before leaves develop. Fruits 4–7.5 cm long, maturing in spring. Often planted as an ornamental, but subject to damage by winds, ice storms, wood rot, and insects. Fruits eaten in abundance by birds and mammals, especially squirrels. Wood of little value. Common. Usually in moist to wet situations, often along streams and margins of lakes. In open or with other hardwoods. Uncommon above 600 m elevation. Feb–Apr.

4.
Mountain Maple
Acer spicatum Lam.
[92]

Identified by stalked buds; by coarsely toothed leaves, teeth about 2–3 per cm. Shrubs or trees to 17 m tall by 35 cm DBH with a brownish, scaly or slightly furrowed bark. Male and female flowers in separate upright clusters. Fruits yellow to red when young, turning brownish at maturity in July–Aug. Browsing animals utilize leaves and twigs, birds and rodents eat the fruits. Occasional. Under larger hardwoods, especially in rich rocky soils. From about 1800 m elevation to rarely below 900 m in sAppalachians; to sea level in neUS and Can. Extends into ceSask, Nfld, and sLab. May–July.

5.
Striped Maple
Acer pensylvanicum L.
[93]

Identified by stalked buds; finely toothed leaves, teeth about 5–10 per cm. Shrubs or trees to 18 m tall by 35 cm DBH. Bark unusual, bright green with conspicuous longitudinal pale to almost white stripes. Flowers in drooping clusters at end of twigs, male and female usually in separate clusters. Fruits widely forking, the wings 1–2 cm long, maturing in late summer or autumn. Occasionally planted as an ornamental, often because of its attractive bark. Birds and rodents eat the fruits and several kinds of animals browse on twigs and leaves. Occasional. Usually an understory tree in rich hardwood forests. Rarely occurs above 1700 m elevation in sAppalachians and uncommon below 750 m; reaching sea level in the ne portion of its range. Extends to nNS. Mar–June.

6.
Norway Maple
Acer platanoides L.
[94]

Recognized by leaves with a few large teeth, but no serrations; by petioles exuding milky juice when broken from twig. Trees to 30 m tall by about 1 m DBH, but usually much smaller. Terminal buds blunt. Flowers in erect rounded clusters, appearing with the leaves. Planted introduction, its dense foliage and rounded crown making an excellent shade tree; leaves bright yellow in autumn. In the SE best adapted to northern sections, does well northward. Rare. Planted throughout most of the US and sCan, often escaping to such places as yard borders, vacant lots, roadsides. Native of Europe and eAsia. Apr–May.

7.
Chalk Maple;
White-bark Maple
Acer leucoderme Small
[95]

Recognized by leaves with only a few large blunt teeth and no serrations, the lobe tip and sides usually drooping, the undersurface yellowish-green and hairy; by multiple trunks (occasionally single) the trunk(s) at least slightly crooked but usually pronounced. Shrubs or trees to 12 m tall and 70 cm DBH. Bark light-colored to almost white. Crown low, rounded. Terminal buds pointed, finely hairy. Flowers appearing with the leaves. Trees hardy, do well in the open, have attributes of an ornamental, but not widely used. Leaves yellow to rich purplish-red in autumn; at least the lower and inner leaves tend to hang on the tree until spring. Fruits sometimes remaining on tree until early autumn; eaten by birds and rodents, especially squirrels. Common. Understory tree in hardwood forests, usually on fairly well-drained stream banks and terraces and on adjacent slopes. Mar–Apr. Syn: *A. saccharum* ssp. *leucoderme* (Small) Rehd.

8.
Black Maple
Acer nigrum Michx. f.
[96]

Distinguished by leaves with only a few large teeth and no serrations, the end and sides of lobes drooping, the undersurface yellowish-green to green and hairy; by having one straight trunk with the bark gray to almost black and becoming deeply furrowed with increased size. Large trees to 35 m tall by about 1.5 m DBH. Terminal buds pointed, finely hairy. Flowers in large clusters from base of developing leaves. Wood of about same quality and uses as the well known *A. saccharum*. Useful source of sap for making maple syrup; planted as an ornamental, being resistant to wind and ice damage. Common. In hardwood forests of lowlands and rich slopes. To over 1600 m elevation in sAppalachians to a little above sea level in the NE. Apr–May.

9.
Sugar Maple
Acer saccharum Marsh.
[97]

Identified by flat leaves with only a few coarse teeth and no serrations, the tip of the lobes sharply acute to acuminate, the margin and lobes not drooping, under-surface green to pale green and glabrous or with hairs on the veins. Trees to 38 m tall by 2 m DBH. Trunk straight; bark light gray to grayish-brown, becoming rough and deeply furrowed; terminal buds pointed. Flowers appearing with the leaves. Important source of wood for flooring, furniture, veneer, gunstocks, and many other lesser uses; wood with curly or wavy patterns especially select. Important source of sap for making maple syrup; useful as a shade tree. Drooping tips and sides of lobes of leaves will separate the similar *A. nigrum*. The similar *A. barbatum* is separated by having lobes of leaf blades rounded to acute and the undersurface always hairy and usually glaucous. Common. Rich well-drained woodlands of level areas or on adjacent slopes. Uncommon above 1600 m elevation, to near sea level in the NE. Extends to nNS. Mar–May.

10.
Southern Sugar Maple
Acer barbatum Michx.
[98]

Identified by leaves with only a few coarse teeth and no serrations, the lobes rounded to acute, the undersurface glaucous and usually hairy. Trees to 30 m tall and about 1 m DBH. Trunk straight; bark resembling that of *A. saccharum;* terminal buds pointed. Flowers appearing with the leaves. Leaves a rich red in autumn; some persisting into late winter. Wood nearly the quality and with similar uses as that of *A. saccharum,* but the supply limited. An excellent shade tree, being resistant to wind and ice damage. In some ways much like *A. saccharum,* which see for separation. Occasional. Rich soils of wooded valleys and adjacent hillsides. A frequent escape from plantings. Mar–May. Syn: *A. floridanum* (Chapm.) Pax.

CORNACEAE: Dogwood Family

4.
CORNUS. Dogwood

Seven species of *Cornus* that attain tree size occur in the SE, but several of these are more commonly shrubs. One of the seven, *C. alternifolia,* belongs to Group J and is described there. All species are distinctive in having pinnately veined leaves with the main lateral veins attached to the basal ⅔ of midrib only and arching in such a manner that the tip of each vein is nearer the midvein than is the middle portion. This pattern of veination may differ below inflorescences and on stunted twigs. Stipules absent. Flowers bisexual, petals 4–5 and separate, the ovary inferior. Fruits are drupes with a thin pulp and 1 or 2 hard stones, each containing a single seed. Fruits are a valuable source of food for wildlife and vegetative parts for forage. There are about 40 species of *Cornus,* mostly of temperate regions of N. Amer, Europe, Asia, and Africa. About 15 species occur in lower US, Can, and Alas.

KEY TO ALL CORNUS SPECIES

1. Leaves alternate (described in Group J)   *C. alternifolia*
1. Leaves opposite   2
   2. Flowers sessile, in dense clusters with 4 large white (rarely pink to red) petallike structures at the base; petals greenish-yellow; fruits in dense clusters, red when mature   1. *C. florida*
   2. Flowers on individual stalks in branched clusters, large petallike structures absent; petals white to cream-colored; fruits scattered in loose clusters   3
      3. Pith of 2-year-old twigs brownish, fruits white   4
         4. Upper surface of leaves without hairs or hairs present and the surface scarcely rough if at all   2. *C. racemosa*
         4. Leaves quite rough to touch with stiff hairs on the upper surface   3. *C. drummondii*
      3. Pith of 2-year-old twigs white, fruits white or blue   5
         5. Plants stoloniferous, older twigs turning gray to brownish, lateral buds stalked, mature fruits dull white   4. *C. stolonifera*
         5. Plants not stoloniferous, older twigs turning red to dark reddish-purple, lateral buds sessile, mature fruits blue   6
           6. Upper surface of leaves with short stiff hairs, mildly rough to touch   5. *C. asperifolia*
           6. Upper surface of leaves glabrous, smooth to touch   6. *C. stricta*

1.
Flowering Dogwood
Cornus florida L.
[99]

Recognized by gray to dark reddish-brown bark, the surface broken into small squares; by flowers with greenish-yellow petals, in terminal dense clusters with 4 large white (rarely pink to red) petallike structures at base; by terminal buttonlike flower buds and clusters of red fruits (as in photograph) in autumn and winter. Trees to 15 m tall by 49 cm DBH. Wood extremely shock resistant, with uses such as mallet heads, jeweler's blocks, small pulleys, tool handles. Exceptional as an ornamental because of its hardiness, moderate size, prominent flower clusters in spring, and red leaves and fruits in autumn. Common. An understory tree or in open; from deep moist soils near streams to well-drained upland areas. Occurs from near sea level to about 1500 m elevation in sAppalachians. Two well-separated populations occur in ne and ceMex. Feb–June.

2.
Gray Dogwood
Cornus racemosa Lam.
[100]

This is one of five opposite-leaved dogwoods having flowers and fruits in branched clusters reported as attaining tree size in the SE. These species will not be described individually as they occur almost entirely as shrubs in this area. Should one of tree dimensions be encountered, the Key to Species will aid in identification and botanical manuals can provide descriptions and habitat information. Principal distributions for the 5 species and the flowering periods in the SE are provided below.

2. **Gray Dogwood.** *Cornus racemosa* **Lam.** From sWVa to Ark; north to Wisc, Mich, sQue; east to cMe and Mass. May–July. Syn: *C. foemina* Mill. ssp. *racemosa* (Lam.) J. S. Wilson.

3. **Roughleaf Dogwood.** *Cornus drummondii* **C. A. Mey.** Miss to csTex; north to seSD and eOhio; east to cKy and Miss. Apr–June.

4. **Red-osier Dogwood.** *Cornus stolonifera* **Michx.** In the SE only in WVa and nVa. Occurs westward to Neb and Col; south to nMex; west to nCal; north to Alas, Ont, and Lab. May–July.

5. **Stiff-cornel Dogwood.** *Cornus asperifolia* **Michx.** Coastal parts of seNC, SC, Ga, and in upper Fla. Apr–June. Syn: *C. microcarpa* Nash; *C. foemina* Mill. ssp. *microcarpa* (Nash) J. S. Wilson.

6. **Swamp Dogwood.** *Cornus stricta* **Lam.** From seVa to SC, Ga, cFla, Ala and to eTex and swInd. Mar–May.

OLEACEAE: Olive Family

5.
OSMANTHUS. Osmanthus

Devilwood
Osmanthus americanus (L.)
Benth. & Hook f. ex
Gray
[101]

Distinguished by leaves pinnately veined, mostly over 5 cm long and 2.5 × as long as wide, evergreen, thick and leathery, margin entire, tip pointed, base acute, and glabrous beneath. Additional identification aids include the appearance of developing axillary clusters of flowers before the fruits have fallen, as may be seen in the photograph, and fruit stalks persisting after fruits have fallen, providing some aspect of the flower-to-fruit cycle on most plants at any time of the year. Shrubs or trees to 11 m tall by over 50 cm DBH with pale bark. Terminal buds lance-shaped, with 2 valvate scales. Corollas white to yellow. Mature fruits dark blue nearly globose drupes 6–14 mm long with one stone. Fruits eaten by birds and small mammals. The wood very hard, tough, and difficult to split; has been used on a limited basis for tool handles and other articles requiring exceptionally strong wood. Common. Dry to moist places in open habitats such as dunes and scrub oak sandhills to rich woodlands. Confined almost entirely to the CP. Feb–May.

6.
FORESTIERA. Forestiera

For identifying characters see individual species. Shrubs or trees with pinnately veined leaves. Flowers usually male and female on separate trees, sometimes bisexual; sepals 4–6, unequal; petals none. Fruits of SE species dark purple to black drupes with thin flesh and a single stone. A genus of about 20 species, 4 native in US, and 2 occurring in the SE.

KEY TO FORESTIERA SPECIES
1. Leaves evergreen, entire, rounded at tip   1. *F. segregata*
1. Leaves deciduous, toothed beyond the middle, pointed at the tip

                                                                              2. *F. acuminata*

1.
Florida-privet
Forestiera segregata (Jacq.)
Krug & Urban
[102]

Recognized by evergreen entire pinnately veined leaves, the vein system prominent on upper surface; by leaves mostly under 5 cm long, glabrous beneath, tip rounded. Shrubs or trees to 8 m tall by 20 cm DBH with smooth grayish-white bark. Flowers in small clusters at leaf bases. Fruits ellipsoid, 6–11 mm long, about 8 mm broad. Rare. Usually in shell-bearing soils: dunes, shell mounds in saltmarshes, hammocks, maritime woods, especially along backside margin of islands. Along Atlantic Coast from Fla to southernmost SC; only on eastern portion of Gulf Coast of Fla. Sept–Apr. Syn: *F. porulosa* (Michx.) Poir.

2.
Swamp-privet
Forestiera acuminata
(Michx.) Poir.
[103]

Identified by deciduous pinnately veined leaves, margin toothed beyond the middle, tip acuminate, stipules absent. Shrubs or trees to 13 m tall by 25 cm DBH with thin smooth brownish bark, often with many lichens and liverworts on the surface. Twigs with prominent lenticels; terminal buds acute; leaf scars with 1 bundle scar. Flowers in small axillary clusters on previous year's twigs, appearing before current year's leaves are developed. Fruits black curved-ovoid drupes 10–15 mm long with thin pulp and a large one-seeded stone. Fruits eaten by ducks. Occasional. Swamps, floodplains, wet stream borders, sandbars, in and at margins of ponds and lakes, wet woods. Mar–May.

7.
CHIONANTHUS. Fringe-tree

Fringe-tree;
Old-man's-beard
Chionanthus virginicus L.
[104]

Recognized by bud scales sharply keeled and with an abrupt sharp point; by stipule scars absent; by leaves deciduous, entire, tip obtuse to short-acuminate. Shrubs or trees to 10 m tall by 33 cm DBH. Bark reddish-brown with small flattened scales on the larger trunks. Twigs usually with prominent warty lenticels. Flowers fragrant, in large loose drooping clusters from previous year's twigs. Note new stems with leaves above flowers in the photograph. Petals 4, white, united at base, the lobes very narrow and to 3 cm long. Fruits dark blue to nearly black drupes 15–25 mm long. Fringe-tree is quite variable. One of the variants reaching tree size in the SE has been named *C. henryae* Li, but does not seem clearly separable from other variants of the species. Fruits eaten by many kinds of wildlife. An attractive ornamental. Occasional. Rock outcrops, dry woods, savannas, hillsides, often as an understory plant. From near sea level to over 1300 m elevation in Appalachian Mts. Another species occurs in China. Mar–June.

8.

LIGUSTRUM. Privet

A variable genus difficult to delineate. For identifying characters refer to Key to Genera of Group E. Shrubs or small trees. Leaves evergreen or deciduous, leaf scars raised, each with one bundle scar. Inflorescence a many-flowered terminal cluster. Petals white to cream-colored, united at base, with 4 lobes. Fruits berrylike drupes with 1–4 seedlike stones. Frequently eaten by birds, and stones distributed in droppings. A genus of about 50 species in Asia and Australia and one in Mediterranean area. None is native to the SE, but at least 10 species have been used ornamentally; most have escaped, some extensively; 5 have been reported to reach tree size in SE. Identification to species is sometimes difficult, the problem enlarged by introduction of horticultural varieties.

KEY TO LIGUSTRUM SPECIES

1. Leaves 7–15 cm long   2
    2. Leaves short-acuminate to nearly obtuse at tip, 7–10 cm long; corolla tube slightly longer than lobes   1. *L. japonicum*
    2. Leaves acuminate at tip, 10–15 cm long; corolla tube equaling lobes
                                               2. *L. lucidum*
1. Leaves 2–6 cm long   3
    3. Twigs glabrous, corolla tube 5–8 mm long   4. *L. ovalifolium*
    3. Twigs short- to minutely hairy, corolla tube 1.5–3.5 mm long   4
        4. Leaves hairy on midrib beneath   3. *L. sinense*
        4. Leaves glabrous beneath   5. *L. vulgare*

1.

Japanese Privet
Ligustrum japonicum
Thunb.
[105]

Identified by glabrous twigs; by evergreen leaves 7–10 cm long, the tip short-acuminate to nearly obtuse, shiny above; by corolla tube slightly longer than the lobes. Shrubs or trees to 12 m tall by 17 cm DBH. Widely used as an ornamental. Rare. Escaped near dwellings and nearby woods, fencerows, and waste places. May–Aug.

2.
Ligustrum lucidum Ait. f.

Similar to *L. japonicum,* but leaves 10–15 cm long, the tip acuminate; the corolla tube equaling the lobes. Shrubs or trees to 10 m tall by 15 cm DBH. Leaves dull above. Used as an ornamental. Rare. Distribution similar to the above species. July.

3.
Chinese Privet
Ligustrum sinense Lour.
[106]

Identified by hairy twigs; by leaves evergreen or some-times deciduous in severe winters or in colder parts of SE, 2.5–6 cm long, hairy on midrib beneath. Shrubs or trees to 8 m tall by 16 cm DBH. Flowers fragrant, corolla tube 1.5–3 mm long. Once widely used as an ornamental. Occasional in some areas, an abundant and obnoxious pest in others, often forming impenetrable thickets, especially in bottomlands and along streams. Escaping from original plantings and spreading, pre-sumably by bird droppings, to woods, fencerows, and waste places. Apr–June.

4.
Ligustrum ovalifolium
Hassk.

Having leaves similar to those of *L. sinense,* but may be recognized by glabrous twigs and a corolla tube 5–8 mm long. Rarely escaping from cultivation and barely attaining tree size in the wild. May–July.

5.
Ligustrum vulgare L.

Having twigs, leaf longevity, and size similar to *L. sinense,* but leaves glabrous beneath. Abundance and distribution similar to that of *L. sinense.* May–June.

VERBENACEAE: Verbena Family

9.
AVICENNIA. Black Mangrove

Black Mangrove
Avicennia germinans (L.) L.
[107]

Recognized by leaves evergreen, pinnately veined, and finely hairy beneath; by fruits flattened capsules. Shrubs or trees, reaching 18 m tall and 70 cm DBH in Florida Everglades, but only occasionally of tree size in the area covered by this book. Bark dark gray to brown, smooth at first, becoming scaly on larger trunks. Plants with many upright unbranched roots that catch debris brought in by the tides. Flowers with a nectar attractive to bees. Rare in general distribution but often in dense colonies that are virtually impen-etrable. Tidal flats in salt and brackish water. Coastal Fla, north on Atlantic to St. Augustine and on Gulf to near Apalachicola; on Miss and eLa islands; s coast of Tex and extending into subtropical and tropical Amer-ica and Africa. About 14 other species cumulatively are widely distributed throughout tropical regions of the world. June–July, occasionally during any part of year. Syn: *A. nitida* Jacq.

BIGNONIACEAE: Bignonia Family

10.
PAULOWNIA. Paulownia

Paulownia;
Princess-tree
Paulownia tomentosa
(Thunb.) Sieb. & Succ. ex
Steud.
[108]

Recognized by leaves palmately veined, unlobed, margin entire, hairy on both sides, densely so beneath; by twigs of previous year hollow except at nodes and occasionally at scattered places between nodes. Trees to 32 m tall by nearly 2 m DBH, but usually much smaller though fast growing. Terminal buds absent, axillary buds sunken in bark; lenticels large, pale, and numerous. Leaves 14–30 cm long, but may reach 60 cm or more on vigorous sprouts when trunks are cut down. Flowers and fruits in large terminal clusters, the flower buds beginning to form in the summer and well developed by autumn, as seen in the photograph. Flowers fragrant, appearing before leaves. Corolla 5–6 cm long, purple, tubular, with 5 short rounded unequal lobes. Fruits ovoid capsules 2.5–4 cm long with 2 carpels and many tiny winged seeds attached to the 2 very large placentas. Often planted as an ornamental for its showy flowers. Common. Frequently escaping, can be troublesome because of abundant reproduction from seeds and vigorous sprouting following cutting. Seeds are blown considerable distances, often establishing sizable colonies along roadsides and stream banks. Frequently escaping in cities and towns and around rural residences. There are about 14 species in the genus; natives of Asia. Apr–May.

11.
CATALPA. Catalpa

Recognized by glabrous twigs, the twigs of previous year with continuous pith; by leaves palmately veined, entire, blades over 10 cm long, and unlobed. Lenticels large, pale, and numerous. Terminal buds absent. Flowers bisexual, the corolla united with 5 lobes, the 3 lower ones smaller than the 2 above. Fruits long narrow cylindrical capsules with many papery-winged seeds. A genus of about 12 species native to N. Amer, W. Indies, and Japan, with 2 species native to the SE attaining tree size. Often planted and spreading far beyond the original natural range.

KEY TO CATALPA SPECIES
1. Leaf tips barely short-acuminate; corolla 2–4 cm broad, the lower lobe not notched; capsules 8–12 mm thick; seed wing gradually narrowed to the end with a narrow tip of hairs   1. *C. bignonioides*
1. Leaf tips acuminate; corolla over 4 cm broad, the lower lobe notched; capsules about 15 mm thick; seed wing rounded at end and with a flat fringe of hairs
                                                                2. *C. speciosa*

1.
Southern Catalpa
Catalpa bignonioides Walt.
[109]

Recognized by barely acuminate leaf tips; by corolla 2–4 cm broad, the lower lobe not notched; by long narrow fruits 8–12 mm thick. Trees to 25 m tall by 1.7 m DBH, flowering when about 7 years old. Planted as an ornamental shade tree and as a source of caterpillars for fish bait. Used for fenceposts and occasionally for furniture and interior work. Occasional. Moist soils, often along streams, sometimes mixed with other broadleaf trees. Original range nFla and swGa to sAla and eMiss, but now widely spread from plantings to Tex, north to Mich and northeast to New England. May–July.

2.
Northern Catalpa
Catalpa speciosa (Warder ex Barney) Engelm.
[110]

Recognized by acuminate leaf tips; by corolla over 4 cm broad, the lower lobe notched; by long narrow fruits about 15 mm thick. About the same size and uses as *C. bignonioides*. Occasional. Usually in moist soils along streams and around lakes, often associated with other broadleaf deciduous trees. Thought to be native from wTenn to neArk, ceIll, and swInd, now widely spread from plantings to eTex, seNeb, Ohio, Md, and SC. May–July.

RUBIACEAE: Madder Family

12.
PINCKNEYA. Pinckneya

Pinckneya; Fever-tree
Pinckneya bracteata (Bartr.) Raf.
[111]

Recognized by leaves deciduous, pinnately veined, hairy on both surfaces; by stipules, or the scars left when they fall, bridging the two spaces between opposite petiole bases; by fruits globose to ovoid 2-celled capsules 12–19 mm broad. Shrubs or trees to 8 m tall by 25 cm DBH. Flowers bisexual, petals greenish-yellow but greatly overshadowed by a conspicuous petal-like creamy to dark pink enlarged sepal on each flower. Extracts of the plant were once considered efficacious for fever, a belief apparently unsupported by medical evidence. Occasionally grown as an ornamental because of its showy inflorescences. Rare. Wet areas of swamps, seepage areas. Occurs as an understory plant in woods, but develops best in partial or complete open. May–July. Syn: *P. pubens* Michx.

13.
CEPHALANTHUS. Buttonbush

Buttonbush
Cephalanthus occidentalis L.
[112]

Identified by opposite or whorled deciduous pinnately veined leaves with blades glabrous, at least on upper side; by stipules, or the scars left when they fall, bridging the spaces between the 2–4 petiole bases at each node; by white corollas; by fruits angular nutlets densely compacted in a globose head. Shrubs or uncommonly trees to 15 m tall by 40 cm DBH with the trunk usually crooked. Terminal buds absent; axillary buds usually well above leaf scar, very small, sunken in depressed areas. Leaf blades entire. Rarely a few leaves persist through winter in the southernmost part of the SE. Corollas 6–12 mm long, united, with 4 lobes; ovary inferior. Sometimes planted as an ornamental. Common. Wet habitats in woods or open. Often abundant on stream banks, lake shores. Usually below 550 m elevation. Besides the distribution shown on the map, the species occurs in parts of Ariz and Cal and much of Mex in wet habitats. May–Sept.

CAPRIFOLIACEAE: Honeysuckle Family

14.
VIBURNUM. Viburnum

Recognized by stipules absent, or present and fastened to petioles, thus no stipule scars on the twigs; by 3 bundle scars in each leaf scar; by absence of sharp keel on bud scales; by deciduous leaves, uncommonly some persisting on 2 species in the southernmost part of the SE; by blades pinnately veined, the lateral veins not prominently arched, the tip short-acuminate to obtuse. Flowers bisexual, in terminal flat-topped or rounded clusters. Corolla white to cream-colored, united, with 5 equal lobes. Ovary inferior. Fruits drupes with a single-seeded stone. Fruits eaten by birds and other wildlife, deer browse vegetation. A genus of about 130 species widespread in temperate and subtropical regions of the world, mostly the N. Hemisphere, but extending south to Java. Six species attain tree size in the SE.

KEY TO VIBURNUM SPECIES

1. Petioles absent or rarely to 5 mm long; leaves mostly widest beyond the middle, tip rounded to obtuse, margin entire or with teeth from about the middle to tip
1. *V. obovatum*
1. Petioles usually over 5 mm long; leaves mostly widest near the middle, tip pointed with some leaves obtuse, margin entire to toothed along the whole margin
2

2. Leaves leathery in texture, shiny above; petioles rusty-hairy; buds densely covered with rusty hairs   2. *V. rufidulum*

2. Leaves neither leathery in texture nor shiny above; petioles and buds without rusty hairs   3

   3. Flower clusters sessile or nearly so; peduncles, if present, under 1 cm long
                                                                                                                                                                   4

      4. Leaves with tiny dark glands beneath   3. *V. lentago*

      4. Leaves without such glands   4. *V. prunifolium*

   3. Flower clusters with peduncles over 1 cm long   5

      5. Leaves crenate, dentate, or undulate, rarely entire, without marginal hairs; base of flower clusters 7–16 mm above first set of leaves
                                                                                                                               5. *V. cassinoides*

      5. Leaves entire to barely undulate, with at least a few marginal hairs (magnification usually needed); base of flower clusters 15–45 mm above first set of leaves after an intervening set of leaves has dropped early
                                                                                                                               6. *V. nudum*

1.
Small-leaf Viburnum
Viburnum obovatum Walt.
[113]

Recognized by leaves sessile or with petioles rarely to 5 mm long; by leaf blades mostly widest beyond the middle, the tip rounded to obtuse, the margin entire or with teeth from about midway to the tip; by flower clusters sessile or on peduncles to 3 mm long. Shrubs or trees to 9 m tall by 18 cm DBH. Leaves 2.5–6 cm long, wedge-shaped or tapered at base; some leaves may remain alive on tree during mild winters. Flowering as leaves develop. Fruits ellipsoid, 6–9 mm long, becoming red and later shiny black. Occasional. Stream banks, floodplains; in dense or thin woods; sometimes in open sandy uplands. Feb–Apr.

2.
Rusty Blackhaw
Viburnum rufidulum Raf.
[114]

Recognized by leathery leaves, rusty-hairy petioles, by blades 4–10 cm long, tip acute to slightly acuminate; by flower clusters sessile or on peduncles to 7 mm long. Shrubs or trees to 9 m tall by 40 cm DBH. Leaf margin serrate. Fruits dark blue and glaucous, ellipsoid, 12–15 cm long. Common. Usually dry habitats, in thin or dense woods. Rarely occurs over 750 m elevation. Mar–May.

3.
Nannyberry
Viburnum lentago L.

Similar to *V. rufidulum* in the character of the leaf tip and margin and in the sessile, or nearly so, flower clusters. It may be separated by the tiny dark glands on the underside of the leaf. Shrubs or trees to 12 m tall by 25 cm DBH. Fruits blue-black and glaucous, ellipsoid to nearly globose, 8–12 mm long. Common. Moist soils in rich woods and along streams; sometimes in rocky uplands. To about 750 m elevation in Appalachians, but to higher elevations outside the SE. May–June.

4.
Blackhaw
Viburnum prunifolium L.

Similar to the above two species, but lacks rusty hairs on the petiole (as in *V. rufidulum*) and tiny dark glands on the underside of leaves (as in *V. lentago*). Shrubs or trees to 10 m tall by 25 cm DBH. Fruits black or sometimes dark blue, occasionally glaucous, ellipsoid, 8–15 mm long. Common. Usually in moist places or rich soils of hillsides. Unusual above 600 m elevation. Mar–May.

5.
Possumhaw Viburnum
Viburnum cassinoides L.
[115]

Recognized by petioles usually over 5 mm long and lacking rusty hairs; by most leaves widest near the middle, the margin crenate, dentate, or undulate, rarely entire; by base of flower clusters 7–16 mm above first set of leaves. Shrubs or trees to 8 m tall by 10 cm DBH. Fruits at first pink to bright red, turning to deep blue or nearly black and glaucous, 5–8 mm long. Common. Wet habitats in woods or open. In sAppalachians most frequently above 1200 m elevation; frequent also at lower elevations northward. Extends to Nfld. Apr–July.

6.
Possumhaw Viburnum
Viburnum nudum L.

Similar in appearance and size to *V. cassinoides*, but may be distinguished by leaf margin entire to barely undulate and the first pair of leaves below flower cluster dropping early, leaving the base of the flower clusters 15–45 mm from the first set of persistent leaves. Common. Wet habitats in woods or in open. Unusual above 600 m elevation. Mar–May.

Group F

Leaves simple, alternate, stipule scars encircling twig or nearly so.

KEY TO GENERA
1. Leaves broadly truncate to broadly notched at apex   2. *Liriodendron*
1. Leaves rounded to acuminate at apex   2
     2. Leaves with three prominent veins arising from one place at base of blade
                        4. *Platanus*
     2. Leaves with one large vein at base of blade   3
         3. Leaf margin sharply toothed   1. *Fagus*
         3. Leaf margin entire   3. *Magnolia*

FAGACEAE: Beech Family

1.
FAGUS. Beech

American Beech
Fagus grandifolia Ehrh.
[116]

Recognized by leaves with dentate margin, the tip acuminate. Trees to 49 m tall by 1.3 m DBH. Bark grayish, smooth on old as well as young trees. Terminal buds narrow, to 25 mm long; axillary buds usually not directly over leaf scar. Leaves 2-ranked, with main lateral veins and marginal teeth equal in number, each vein ending in a tooth. Stipules and stipule scars almost but not quite encircling twig. Dead leaves often persistent during winter. Flowers male and female, both on the same tree. Fruits 3-angled nuts, usually 2 in 4-parted burs covered with spines. The nuts are an important source of food for many species of wildlife, including birds, several rodents, and bears. The wood is hard and strong, but not durable and with few uses, mainly flooring, furniture, and tool handles. Common. Rich soils of uplands, well-drained soils of lowlands, sometimes in moist soils. From near sea level to over 1800 m elevation in sAppalachians. Extends to nNS. Also occurs in ec mountains of Mex. Nine other species occur in the cooler parts of the N. Hemisphere. Feb–May.

MAGNOLIACEAE: Magnolia Family

2.
LIRIODENDRON. Yellow-poplar

Yellow-poplar;
Tuliptree
Liriodendron tulipifera L.
[117]

Recognized by leaves broadly truncate to broadly notched at apex and with 2 or 4 lateral lobes and 2 terminal lobes. Trees to 38 m tall by 2.9 m DBH, with dark gray bark that becomes thick and deeply furrowed on older trees. Twigs with bitter taste. Winter buds a little flattened, rounded at the tip, covered by 2 stipules that meet at their edges. Pith continuous, but with firm diaphragms. Sepals 3, reflexed; petals 6, in 2 rows. Fruits single-seeded samaras, these arranged tightly in oblong pointed cones to about 7 cm long. Wood soft and lightweight, the brownish heartwood quite durable, having many uses including furniture, interior finishes, plywood, musical instruments, and pulpwood. Common. Usually most abundant in moist rich soils, but easily moves into drier sites such as abandoned fields and timbered uplands. From near sea level to over 1400 m elevation in sAppalachians. Another species occurs in temperate eAsia. Mar–June.

3.
MAGNOLIA. Magnolia

Recognized by leaf margin entire, some species having blades without lobes and some with base eared or heart-shaped. Trees or shrubs. Pith continuous, but often with firm diaphragms. Buds covered by two fused stipules. Perianth of 9–18 segments, the sepals usually difficult to distinguish from petals. Each flower with stamens and pistils, the latter obviously arranged spirally in a cone-shaped structure. Fruits conelike, an aggregate of leathery follicles, each formed from a single pistil and bearing 1 or 2 seeds that, when released, usually hang for some time on a long threadlike stalk. Seeds eaten by wild animals, especially birds. A genus of about 75 species, about 50 in seAsia, 25 from Venezuela to eUS; 8 occur in the SE.

KEY TO MAGNOLIA SPECIES
1. Leaves acuminate to slightly cordate at base   2
   2. Leaves glaucous beneath, aromatic, petals usually less than 5 cm long
                                 1. *M. virginiana*
   2. Leaves green or rusty beneath, not aromatic, petals usually more than 5 cm
     long   3
     3. Leaves evergreen, coriaceous   2. *M. grandiflora*
     3. Leaves deciduous, membranous   4

4. Most leaves clustered terminally on twig   3. *M. tripetala*
4. Leaves scattered along twig   4. *M. acuminata*
1. Leaves deeply cordate or eared at base   5
   5. Buds and twigs hairy, leaves white beneath   6
      6. Trunk upright, stamens 15–18 mm long, conelike fruits subglobose to globose   5. *M. macrophylla*
      6. Trunk usually not strictly upright, stamens 11–13 mm long, conelike fruits subcylindric to ovoid   6. *M. ashei*
   5. Buds and twigs glabrous, leaves green below   7
      7. Stamens 8–15 mm long, most leaves over 25 cm long, conelike fruits 6.5–11 cm long   7. *M. fraseri*
      7. Stamens 4–6 mm long, most leaves less than 25 cm long, conelike fruits 3.5–5.5(6) cm long   8. *M. pyramidata*

1.
Sweetbay; Swampbay
Magnolia virginiana L.
[118]

Identified by leaves acute to obtuse at base, the underside glaucous. Trees to 27 m tall by about 1.2 m DBH, or often shrubs. Bark gray, smooth, thin. Bruised leaves, twigs, and bark aromatic. Terminal buds with fine silky light-colored hairs. Some or most leaves remain alive through winter in the southern parts of its range, but most or all leaves fall before spring in the coldest portions. Flowers fragrant, with 8–12 petals 25–45 mm long. Fruits 25–55 mm long, ellipsoid to subglobose, the seeds red. Wood used for pulp and sparingly for such things as furniture, tool handles, and interior finishing. Plants easily recognized from a distance when breezes expose the light-colored undersurface of leaves. Common. Wet places in woods or in open; along streams; in flatwoods and swamps. To about 200 m elevation. Mar–July.

2.
Southern Magnolia; Bullbay
Magnolia grandiflora L.
[119]

Recognized by coriaceous evergreen leaves with underside bearing rust-colored hairs (rarely these may be absent and the undersurface green), with bases acute to obtuse. Trees to 30 m tall by 1.8 m DBH. Twigs at first covered with rust-colored hairs, but becoming smooth with age. Petals 8–12 cm long. Conelike fruits oblong to ellipsoid, 6–10 cm long. Seeds scarlet; eaten by birds and rodents. Wood hard and heavy; used for pulp and sparingly for furniture and interior work. Widely planted as an ornamental, frequently beyond its natural range; sometimes adventive near such plantings. Common. Lowland woods, wooded ravines, swamp margins, hammocks, maritime woods, dunes (where it may appear to be shrubby due to burial of lower parts by sand). Occurs to about 165 m elevation, a few escapes from plantings to 250 m elevation. Mar–June.

3.
Umbrella Magnolia
Magnolia tripetala L.
[120]

Recognized by leaves deciduous and clustered terminally on twigs in umbrella fashion, blades 10–45 cm long with the base long acuminate. Trees to 25 m tall by over 1 m DBH, usually much smaller, sometimes with more than one trunk. Petals white, 8–12 cm long. Fruits 6–15 cm long. Of little commercial value; sometimes planted as an ornamental. Seeds eaten by wildlife. Occasional. Rich hardwood forests, often in moist places. To 1065 m elevation in sAppalachians, but usually below 750 m elevation. Apr–May.

4.
Cucumbertree
Magnolia acuminata (L.) L.
[121]

Recognized by leaves scattered along twig, deciduous, the blades 10–30 cm long, pale green beneath, the base obtuse, or some rounded to somewhat cordate. Trees to 30 m tall by 1.8 m DBH. Leaves normally with soft hairs beneath, the tip obtuse to acuminate. Petals 4–8 cm long, greenish-yellow to yellow. Fruits 3–8 cm long, often irregularly shaped. Usually separated into two varieties: var. *acuminata* with twigs of current year glabrous, perianth segments greenish-yellow or sometimes the upper surface yellow; and var. *subcordata* (Spach) Dandy, once considered a separate species, *M. cordata* Michx., with twigs of the current year having short hairs or roughened by hair bases, perianth segments light yellow on underside and golden yellow above, rarely greenish-yellow. The latter variety occurs mostly in the CP and Piedmont; var. *acuminata* occurs in the interior, especially in the mountains. The photograph is of var. *subcordata.* In the Ark and Mo mountains there are plants having lower leaf surface green and glabrous or only slightly hairy on the veins, instead of pale green and hairy as in others, and these have been separated as var. *ozarkensis* Ashe. The differences do not seem to be sufficient to merit recognition of the variety. Wood weak, used sparingly for such things as furniture, cabinets, crating, pulpwood. Seeds eaten by wildlife. Occasionally planted as an ornamental. Common. Rich wooded slopes, upland coves, stream banks. From about 30 m elevation near the Gulf to about 1500 m in sAppalachians, but rarely common above 900 m. Mar–June.

5.
Bigleaf Magnolia
Magnolia macrophylla
Michx.
[122]

Recognized by upright trunk; by leaves deciduous, the blades to 1 m long and 30 cm wide, rounded to lobed at base, lower surface white and with fine silvery hairs; by stamens 15–18 mm long. Trees to 18 m tall by 90 cm DBH. Twigs with silvery hairs when young, the hairs usually falling by the end of the first growing season. Petals 15–20 cm long, white to cream-colored. Fruits subglobose to globose, 6–11 cm long. Of minor use because of its scarcity in timber size. Attractive as an ornamental, but not very practical as large leaves are easily damaged by wind. Occasional. Bottomland woods, wooded ravines, rich wooded slopes. Apr–June.

6.
Ashe Magnolia
Magnolia ashei Weatherby

Similar to *M. macrophylla,* but the trunk usually not upright, stamens 11–13 mm long, and the conelike fruits subcylindric to ovoid. Shrubs, rarely trees to 10.5 m tall by 15 cm DBH, branching near the base. Plants flowering when very small; one only 10 cm tall with a flower has been recorded. Rare. In woods, slopes, bluffs, hills, ravines, and watercourses. In nwFla only. Mar–Apr.

7.
Fraser Magnolia
Magnolia fraseri Walt.
[123]

Recognized by leaves deciduous, lobed at base, lower surface green and glabrous; by stamens 8–15 mm long; by conelike fruits 6.5–11 cm long. Trees to 24 m tall by 90 cm DBH. Leaves mostly crowded near end of current year's growth. Buds glabrous. Flowers fragrant; petals white to cream-colored, 8–13 cm long. Trees of size for lumber scarce, wood soft and weak. Wildlife eat the seeds. Common. Rich woods; along creeks and rivers, and especially on mountain slopes. Present to over 1500 m elevation in sAppalachians. Mar–June.

8.
Pyramid Magnolia
Magnolia pyramidata
Bartr.
[124]

Recognized by leaves deciduous, lobed at base, lower surface green and glabrous; by stamens 4–6 mm long; by conelike fruits 3.5–5.5, rarely to 6, cm long. Trees to 17 m tall by 60 cm DBH. Petals white to creamy-white, 5–11 cm long. This species sometimes occurs in the same area as *M. ashei,* which can be distinguished by leaves white-hairy beneath whereas those of *M. pyramidata* are green and glabrous. Of little commercial value because of scarcity. Rare. Dense rich woods on bluffs, slopes, and uplands. Few trees occur over 100 m elevation. Mar–May.

PLATANACEAE: Sycamore Family

4.
PLATANUS. Sycamore

American Sycamore
Platanus occidentalis L.
[125]

Recognized by leaves with 3 prominent veins arising from one place at base of blade, blades with shallow, acute lobes, or these rarely rounded. Trees to 35 m tall by 3.8 m DBH. Bark on young trees with small dark scales. Bark at base of large trunks dark brown and deeply furrowed; bark on upper part of large trunks peeling away in thin scales, often large, exposing a lighter-colored, often nearly white, inner bark. Terminal buds absent; lateral buds hidden under leaf petioles until leaves fall, with one resinous, caplike scale. Flowers tiny, male and female separate on same tree, in dense globose heads that hang down. Male heads about 1 cm across, the female 2–3.5 cm across when in fruit. Trees fast-growing. The wood is hard and coarse-grained, but not very strong. Uses include furniture, interior trim, particle board, pulpwood, fiberboard. Common. Wet places; swamps, edges of lakes, along streams, in bottomlands, along drainageways; a pioneer tree in open upland sites such as old fields, roadsides. To 975 m elevation in sAppalachians. Extends into mountains of neMex. Two other species occur in the wUS and south into wMex. Six species occur in Eurasia. Mar–June.

Group G

Leaves simple, alternate, palmately veined.

14. Margin of leaf blades doubly serrate or with serrated dentations
    (See under Group I)   *Betula*
14. Margin simply serrated or crenate   15
    15. Three main basal veins ending at or near tip of leaf blade
    (rare escape)   9. *Ziziphus*
    15. Three main basal veins ending in three separate sections of
    the blade   16
        16. Leaves often lobed; juice of twigs, if any, cloudy or
        milky; mature buds with 5–6 exposed scales,
        symmetrical at base, the first scale about ¼ as long as
        bud   3. *Morus*
        16. Leaves never lobed; any juice clear; mature buds with 1–
        2 exposed scales, oblique at base, the first scale about ½
        as long as bud   10. *Tilia*
12. Leaves more than 2-ranked   17
  17. Foliage, buds, and twigs spicy-aromatic when crushed or bruised   18
    18. Leaf margin serrate (rare escape in lower CP—see reference under
        Populus)   *Hovenia*
    18. Leaf margin entire   19
        19. Leaves with yellowish calluslike growths in principal vein
        angles on the upper side of blades; blades all unlobed (see
        Group H)   *Cinnamomum*
        19. Leaves lacking callosities in vein angles on upper side of
        blades; blades lobed or unlobed or both types on same tree
                                                                 4. *Sassafras*
  17. Foliage, buds, and twigs not spicy-aromatic   20
    20. Leaves markedly scabrous above and velvety hairy below, often
        some leaves opposite (See under Group E)   *Broussonetia*
    20. Leaves neither markedly scabrous above nor velvety below, all
        leaves alternate   21
        21. Current year's twigs with stellate hairs in lines or sometimes
        all over the entire surface (uncommon escape)   11. *Hibiscus*
        21. Current year's twigs glabrous except when young, or with
        nonstellate hairs   22
            22. End buds true terminal buds   1. *Populus*
            22. End buds axillary (See under Group F)   *Platanus*

SALICACEAE: Willow Family

1.

POPULUS. Poplar; Cottonwood; Aspen

Recognized by twigs glabrous or with simple hairs; by the end bud a true terminal
bud, winter buds with more than one exposed scale, the lateral buds with the lowest
scale placed directly above the leaf scar; by leaves more than 2-ranked, unlobed,
with small to large, sometimes rounded, teeth; by foliage and twigs not spicy-
aromatic when bruised. Southeastern species are thornless trees with leaves usually

less than 3 × as long as wide. Stipules present. Male and female flowers tiny, crowded in narrow catkins, the sexes on separate trees. Flowers appearing in early spring before or as the leaves appear. Fruits small capsules containing many tiny seeds, each bearing a tuft of cottony hairs, thus easily airborne. Seeds may be blown great distances and can clog window screens and air filters. Wood light-colored, soft, not durable in soil; has uses such as pulpwood, plywood, low quality lumber, veneer, and matchsticks. Buds and flower clusters eaten by several species of birds; leaves, buds, and twigs eaten by various other animals. A genus of about 40 species occurring in Eurasia and N. Amer, being most abundant in the colder regions where some species are low shrubs, especially in the tundra. There are 7 tree species in the SE, some of these extending far beyond this area. Four other species occur in wUS and extend into Can or Mex, or both. Hybrids of *Populus* spp. have been planted in various localities in the SE and some are known to have escaped. Only one of these is included, as the others are rarely encountered.

Hovenia dulcis Thunb., Japanese Raisin-tree (Rhamnaceae), is a rare tree with all the identifying recognition characteristics of *Populus* except that the end bud is axillary rather than terminal. It is a native of China that has escaped from cultivation in Wake Co., NC, and possibly elsewhere. May.

KEY TO POPULUS SPECIES

1. Outer end of petioles conspicuously flattened   2
    2. Leaves with a narrow translucent border   3
        3. Small glands on upper side of base of most blades at junction with petiole; crown not columnar   1. *P. deltoides*
        3. Such glands absent; crown usually nearly columnar   2. *P. nigra*
    2. Leaves lacking translucent border   4
        4. Leaves with 5–18 teeth on a side; buds dull brown, finely hairy   3. *P. grandidentata*
        4. Leaves with 20–40 teeth on a side; buds shiny, glabrous, or lowest scales ciliate   4. *P. tremuloides*
1. Outer end of petioles terete or nearly so   5
    5. Leaves white-felted or gray hairy beneath, teeth less than 15 on each side   5. *P. alba*
    5. Leaves glabrous beneath, or with other types of hairs, teeth more than 15 on each side   6
        6. Terminal buds sticky to touch, with an odor of balsam   6. *P. balsamifera*
        6. Terminal buds not sticky to touch, without odor of balsam   7. *P. heterophylla*

1.
Eastern Cottonwood
Populus deltoides Bartr. ex
Marsh.
[126]

Recognized by portion of petiole adjacent to leaf blade conspicuously flattened perpendicular to the blade; by all or most blades bearing small glands at junction with the petiole; by leaves with a narrow translucent border. Rapidly growing trees to 47 m tall by 2.6 m DBH with a broad crown. Flowers in catkins 4–6 cm

long, the female catkins enlarging in fruit to 15–25 cm long. Common. Swamps, along streams, moist lowlands, seepage slopes. Generally absent from sAppalachians. Rarely over 300 m elevation in the eUS, but occurs to above 1500 m in some localities in wUS. Extends into nTex, eCol, eWyo, sAlta, and sSask. Feb–May.

2.
Lombardy Poplar;
Black Poplar
Populus nigra L.
[127]

Recognized by portion of petiole adjacent to leaf blade conspicuously flattened perpendicular to the blade; by absence of glands on upper side of blade at junction with petiole; by leaves with a narrow translucent border. Trees to 30 m tall by 60 cm DBH. All plants are male, reproducing vegetatively. Native of Eurasia, but introduced as an ornamental and escaping. By far the most common form is var. *italica* Moench., Lombardy Poplar, which has a narrow columnar crown with upright branches. Forms with broad crowns, known as Black Poplar, are seldom planted. Rare. Escaping and spreading by root sprouts at abandoned homesites, in fencerows, and other available areas near plantings. Mar–May.

3.
Bigtooth Aspen
Populus grandidentata
Michx.
[128]

Recognized by portion of petiole adjacent to leaf blade conspicuously flattened perpendicular to the blade; by leaves with 5–18 coarse teeth on a side. Trees to 28 m tall by 1.7 m DBH, usually much smaller. Bud scales brown with grayish downy hairs. Well-developed twigs with orange lenticels. Female catkins 7–12 cm long when mature. Reproduces abundantly by root suckers as well as by seeds. Often found with *P. tremuloides*. This species is one of the pioneers following fire and logging and in abandoned fields but, being highly intolerant of shading, is soon replaced by other tree species. Wood used largely for pulp; vegetative parts consumed by wildlife; a favorite of beaver, who use it for food and dams. Common. Usually in dry habitats; thin woods, sandy upland soils, less frequently along streams and at margins of lakes and swamps. To over 900 m elevation in sAppalachians. Extends to nNS. Mar–May.

4.
Quaking Aspen
Populus tremuloides Michx.
[129]

Recognized by portion of petiole adjacent to leaf blade conspicuously flattened perpendicular to the blade; by leaves with 20–40 crenate to serrate teeth on a side. Trees to 33 m tall by 87 cm DBH. Winter buds sharp-tipped with sticky bud scales. Seed-bearing catkins about 10 cm long. Reproducing by root suckers as well as by seeds; a pioneer in a manner similar to *P. grandidentata* and often found with it. Utilization by wildlife also similar. Wood used for such purposes as pulp, particle-board, boxes, and excelsior. Rare. Occurring in a variety of soils in the open or in thin woods. Scattered localities in the SE, mostly at higher elevations. Common north to Lab and Alas, south to the mountains of wUS; also in scattered localities in high mountains of Mex. Apr–May.

5.
White Poplar
Populus alba L.
[130]

Recognized by leaves white-felted hairy beneath, teeth on margin less than 15 on each side. Trees to over 20 m tall by 60 cm DBH. Leaves sometimes with lobes similar to those of some maples. Both sexes rarely sufficiently close to each other for seed production. Planted as an ornamental, but can become obnoxious as it sprouts profusely from roots, frequently forming dense colonies. *P. alba* var. *canescens* Ait. [*P.* × *canescens* (Ait) J. D. Sm.], Gray Poplar, has leaves gray-hairy beneath, becoming nearly glabrous with age, and no lobes. Occasional. In scattered localities where it has spread from nearby plantings, especially at abandoned homesites. Native of Eurasia. Mar–May.

6.
Balsam Poplar
Populus balsamifera L.
[131]

Recognized by leaves glabrous or nearly so beneath; by portion of petiole adjacent to blade terete or nearly so; by terminal buds sticky and with an odor of balsam. Trees to 39 m tall by 1.2 m DBH. Leaves with small blunt teeth, often with resin blotches on undersurface. Female catkins 10–13 cm long when mature. Frequently sprouting from roots; often in pure stands. The wood light, used mostly for pulp and relatively low-quality lumber. Rare. In thin woods or open; intolerant of shading. Common over large areas outside the SE, occurring in n half of Rocky Mts. to Alas and Lab. There are hybrids with other poplars and they usually have sticky buds. Some of these hybrids, named *P. candicans* Ait. or *P. gileadensis* Rouleau and commonly called Balm-of-Gilead, have been planted as ornamentals and have spread by root sprouts in scattered localities. Apr–May.

7.
Swamp Cottonwood
Populus heterophylla L.
[132]

Identified by portion of petiole adjacent to blade terete or nearly so; by teeth on margin of leaves more than 15 on each side; by terminal buds not sticky to touch. Trees to 35 m tall by 1.9 m DBH. Pith of 2-year-old twigs orange. Leaf blades to 25 cm long, the tips rounded to obtuse, the margin with small crenate to serrate teeth, the underside at first whitish hairy, later nearly glabrous. Plants fast-growing, the wood used for pulp and relatively low quality lumber. Occasional, sometimes locally common. Moist to wet soils; river-bottoms, swamps, edges of sloughs. Mar–May.

ULMACEAE: Elm Family

2.
CELTIS. Hackberry; Sugarberry

Recognized by leaves 2-ranked, blades with 3 prominent veins arising from one place at base; by pith of mature twigs with closely and usually irregularly spaced chambers separated by soft partitions, sometimes almost continuously hollow. Shrubs or trees. Buds small, end buds axillary ones; axillary buds appressed. Leaf blades entire to toothed, the sides unequal. Flowers appearing with the leaves. Flowers functionally unisexual, both sexes on same plant; female flowers with a single pistil and 4–5 infertile stamens, male flowers with 4–5 stamens. Fruits globose glabrous nearly dry drupes, the stone with a single seed, the flesh of SE species usually sweetish to the taste, perhaps the reason that some *Celtis* species are called Sugarberry. Some species vary considerably in form and leaf characteristics and, as a consequence, the number of species, subspecies, and varieties varies with authors, making identification of individual specimens difficult. A genus of about 65 species widely distributed in tropical and northern temperate parts of the world. Six species occur in the US, one extending into Can and 3 into Mex; 3 species of tree size occur in the SE.

KEY TO CELTIS SPECIES
1. Leaf blades over 2 × as long as wide, yellowish-green on both surfaces, margin entire or with a few or rarely many teeth   1. *C. laevigata*
1. Leaf blades less than to a little over 2 × as long as wide, dark green above, paler beneath, margin entire to prominently serrate   2

2. Leaf blades 6–17 cm long, around 2 × as long as wide, margin sharply serrate except at base, apex long-acuminate and mostly falcate; mature fruits orange-red to dark purple, dried ripe fruits deeply puckered   2. *C. occidentalis*

2. Leaf blades 2–9 cm long, under 2 × as long as wide, margin entire or, if sharply toothed, entire at base, apex acuminate and rarely falcate; mature fruits light-orange to red or brown, dried ripe fruits smooth   3. *C. tenuifolia*

1.
Sugarberry; Hackberry
Celtis laevigata Willd.
[133]

Recognized by most leaf blades well over 2 × as long as wide, yellowish-green above and below, margin entire or with a few to rarely many teeth. Trees to 28 m tall by 2.4 m DBH. Bark usually smooth when young, later with corky outgrowths which may be scattered to dense, the spaces between smooth. Twigs at first greenish; late in year they may turn brown. Leaves usually only slightly oblique at base, tapering to a long-pointed tip. Fruits 5–8 mm across, orange-red or less commonly purple, the pulp sweet, eaten by birds. Planted as an ornamental, mostly as a street tree, but branches are easily broken by wind and ice. The soft wood has limited use for such items as plywood, furniture, and pulp. Common. River bottomlands, floodplains, margins of sloughs, usually with other hardwoods, but sometimes in pure stands; occasionally in uplands, roadsides, fencerows, in woods. To about 600 feet elevation. Extending in scattered localities into wTex and neMex. Mar–May. Syn: *C. mississippiensis* Spach.

2.
Hackberry
Celtis occidentalis L.
[134]

Identified by leaf blades 6–17 cm long, less than to a little over 2 × as long as wide, dark green above, paler below, the margin serrate, apex long-acuminate and mostly falcate; by dried ripe fruits deeply puckered. Trees to 35 m tall by 1.8 m DBH. Bark at first smooth or nearly so, later becoming spotted to nearly covered with conspicuous corky wartlike projections or with corky ridges as in the photograph. Leaf size and shape are quite variable, leading to the naming of varieties; however, intergradation is so prevalent that probably no varieties should be maintained. Fruits orange-red to dark purple; eaten by many bird species. Wood soft and weak, used for pulp and providing low quality products, such as plywood, crating, inexpensive furniture. Plants resistant to drought; used as street trees in areas of low rainfall. Common. Grows best in bottomlands and other moist soils; also grows on slopes, bluffs, limestone outcrops and soils. Apr–May.

1] *Torreya taxifolia* × ³⁄₇

[2] *Taxus* sp. × ⅔

3] *Pinus strobus* × ⅙

[4] *Pinus palustris* × ⅚

[5] *Pinus elliottii* × ⅕

[6] *Pinus taeda* × ⅕

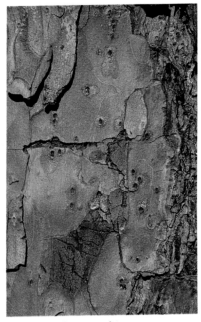

[7] *Pinus echinata* × ⅗

[8] *Pinus echinata* × ¼

[9] *Pinus pungens* × ²⁄₇

[10] *Pinus virginiana* × ¼

[11] *Pinus glabra* × ⅕

[12] *Pinus clausa* × ⅓

[13] *Pinus rigida* × ¼

[14] *Pinus serotina* × ⅕

[15] *Pinus resinosa* × ¼

[16] *Cedrus deodara* × ¼

[17] *Larix laricina* × ⅔

[18] *Picea rubens* × ⅔

[19] *Tsuga canadensis* × ⅞

[20] *Abies fraseri* × ⅓

[21] *Abies balsamea* × ½

[22] *Cunninghamia lanceolata* × ¼

[23] *Taxodium distichum* × ⅓

[24] *Taxodium ascendens* × ⅓

[25] *Thuja occidentalis* × ⁷⁄₁₀

[26] *Chamaecyparis thyoides* × ³⁄₅

[27] *Juniperus communis* × ½

[28] *Juniperus virginiana* × ³⁄₅

[29] *Tamarix canariensis* × ¾

[30] *Tamarix ramosissima* × ¾

[31] *Sabal palmetto* × ¹⁄₆₄

[32] *Serenoa repens* × ¹⁄₇₀

[33] *Yucca gloriosa* × ¹⁄₁₆

[34] *Yucca gloriosa* × 1²⁄₅

[35] *Yucca aloifolia* × ¹⁄₁₀

[36] *Staphylea trifolia* × ⅖

[37] *Acer negundo* × ⅕

[38] *Acer negundo* × ⅕

[39] *Aesculus parviflora* × ¹⁄₃₀

[40] *Aesculus glabra* × ³⁄₁₀

[41] *Aesculus pavia* × ²⁄₉

[42] *Aesculus sylvatica* × ⅐

[43] *Aesculus flava* × ⅑

[44] *Aesculus hippocastanum* × ¼

[45] *Fraxinus americana* × ²⁄₇

[46] *Fraxinus quadrangulata* × 1¼

[48] *Fraxinus profunda* × 1

[47] *Fraxinus pennsylvanica* × 1¼

[49] *Fraxinus caroliniana* × ¼

[50] *Vitex agnus-castus* × ⅓

[51] *Sambucus canadensis* × ⅐

[52] *Juglans nigra* × ⅑

[53] *Juglans nigra* × ⅞

[54] *Juglans cinerea* × ⅚

[55] *Carya illinoensis* × ¹/₇

[56] *Carya myristiciformis* × ¹/₅

[57] *Carya cordiformis* × ²/₅

[58] *Carya aquatica* × ⅙

[59] *Carya ovata* × ⅖

[60] *Carya laciniosa* × ⅙

[61] *Carya tomentosa* × 1

[62] *Carya pallida* × ⅑

[63] *Carya glabra* × ⅔

[64] *Carya texana* × ⅙

[65] *Sorbus americana* × ⅖

[66] *Albizia julibrissin* × ⅓

[67] *Prosopis glandulosus* × ½

[68] *Gleditsia triacanthos* × ⅐

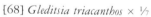

[69] *Gymnocladus dioicus* × ⅛

[70] *Cladrastis kentukea* × ½

[71] *Robinia pseudoacacia* × ¼

[72] *Robinia hispida* × ⅕

[73] *Sophora affinis* × ⅕

[74] *Zanthoxylum americanum* × ⅓

[75] *Zanthoxylum clava-herculis* × ⅓

[76] *Ptelea trifoliata* × ¼

[77] *Poncirus trifoliata* × ³⁄₁₀

[78] *Ailanthus altissima* × ¼

[79] *Melia azedarach* × ⅛

[80] *Rhus glabra* × ¹⁄₁₂

[81] *Rhus copallina* × ⅕

[82] *Rhus typhina* × ²⁄₇

[83] *Toxicodendron vernix* × ⅛

[84] *Sapindus drummondii* × ⅕

[85] *Koelreuteria paniculata* × ⅛

[86] *Aralia spinosa* × 1/15

E

[87] *Broussonetia papyrifera* × ²/₇

[88] *Euonymus atropurpureus* × ⅓

9] *Acer rubrum* × ½

[90] *Acer rubrum* × ⅓

[91] *Acer saccharinum* × ²⁄₉

[92] *Acer spicatum* × ¼

[93] *Acer pensylvanicum* × ¼

[94] *Acer platanoides* × ⅙

E

[96] *Acer nigrum* × ¼

[95] *Acer leucoderme* × ⅓

[98] *Acer barbatum* × ²⁄₉

[97] *Acer saccharum* × ⅓

[99] *Cornus florida* × ⅓

[100] *Cornus racemosa* × ½

[101] *Osmanthus americanus* × ¼

[102] *Forestiera segregata* × ⅔

[103] *Forestiera acuminata* × ²/₇

[104] *Chionanthus virginicus* × ¹/₁₄

105] *Ligustrum japonicum* × ²/₇

[106] *Ligustrum sinense* × ²/₇

[108] *Paulownia tomentosa* × ⅟₇

[107] *Avicennia germinans* × ½

[109] *Catalpa bignonioides* × ⅟₇

[110] *Catalpa speciosa* × ³⁄₁₀

[111] *Pinckneya bracteata* × ¼

[112] *Cephalanthus occidentalis* × ¹⁄₁₄

[113] *Viburnum obovatum* × ⁴⁄₉

[114] *Viburnum rufidulum* × ²⁄₇

[115] *Viburnum cassinoides* × ²⁄₅

F

[116] *Fagus grandifolia* × ³⁄₁₀

[118] *Magnolia virginiana* × ²⁄₃

[117] *Liriodendron tulipifera* × ¹⁄₃

[119] *Magnolia grandiflora* × ⅓

[120] *Magnolia tripetala* × ⅓

[122] *Magnolia macrophylla* × ⅛

[121] *Magnolia acuminata* × ½

[123] *Magnolia fraseri* × ¹⁄₁₅

[124] *Magnolia pyramidata* × ²⁄₇

[125] *Platanus occidentalis* × ¼

[126] *Populus deltoides* × ⅕

[127] *Populus nigra* × ³⁄₁₀

[128] *Populus grandidentata* × ²⁄₉

[129] *Populus tremuloides* × ⅜

[130] *Populus alba* × ³⁄₁₀

[131] *Populus balsamifera* × ¼

[132] *Populus heterophylla* × ²⁄₉

[133] *Celtis laevigata* × ³⁄₁₀

[134] *Celtis occidentalis* × ¹⁄₇

[135] *Celtis tenuifolia* × ¹⁄₄

{136} *Morus rubra* × ²⁄₉

{137} *Morus alba* × ¼

138] *Sassafras albidum* × ²⁄₉

{139} *Liquidambar styraciflua* × ½

[140] *Cercis canadensis* × ⅚

[141] *Aleurites fordii* × ¼

[142] *Manihot grahamii* × ⅙

STERCULIACEAE: Sterculia Family 159

[143] *Ziziphus jujuba* × ⅓

[144] *Tilia americana* × ¼

[145] *Hibiscus syriacus* × 2/7

[146] *Firmiana simplex* × 1/13

[147] *Myrica cerifera* × ½

[148] *Myrica inodora* × ⅜

H

[149] *Quercus virginiana* × ½

[150] *Quercus geminata* × ⅓

[151] *Quercus myrtifolia* × 9/10

{152} *Quercus laurifolia* × ¼

{153} *Quercus hemisphaerica* × ³⁄₇

{154} *Illicium floridanum* × 1¼

{155} *Illicium parviflorum* × ²⁄₉

156] *Cinnamomum camphora* × ⅓

[157] *Persea borbonia* × ⅜

[158] *Cliftonia monophylla* × ⁷⁄₁₀

[159] *Cyrilla racemiflora* × ¼

{160} *Ilex opaca* × ¼

{161} *Ilex vomitoria* × ⅔

{162} *Ilex coriacea* × ⅓

{163} *Ilex cassine* × 2/7

[164] *Ilex myrtifolia* × ⅔

[165] *Gordonia lasianthus* × ¼

[166] *Rhododendron maximum* × ⅕

[167] *Rhododendron catawbiense* × ¹⁄₇

[168] *Kalmia latifolia* × ⅓

[170] *Vaccinium arboreum* × ⅓

[169] *Lyonia ferruginea* × ⅖

[171] *Bumelia lycioides* × ¼

[172] *Bumelia tenax* × ½

[173] *Symplocos tinctoria* × ⅓

[174] *Baccharis halimifolia* × ¼

[175] *Carpinus caroliniana* × ⅕

[176] *Ostrya virginiana* × ½

[177] *Betula alleghaniensis* × ⅖

[178] *Betula lenta* × ⅜

[179] *Betula papyrifera* × ¹⁄₁₆₅

[180] *Betula nigra* × ¼

[181] *Castanea pumila* × ³⁄₇

[182] *Castanea mollissima* × ¼

[183] *Castanea dentata* × ⅕

[184] *Ulmus parvifolia* × ⅗

[185] *Ulmus rubra* × 1½

[186] *Ulmus americana* × 2/7

[187] *Ulmus alata* × ⅘

[188] *Ulmus thomasii* × 1¼

[189] *Ulmus serotina* × 1⅗

[190] *Planera aquatica* × ⅔

[191] *Zelkova serrata* × ⅗

[192] *Asimina parviflora* × ⅜

[193] *Asimina triloba* × 1¹⁄₁₀

[194] *Hamamelis virginiana* × ⅓

[195] *Amelanchier arborea* × ½

{196} *Stewartia malacodendron* × 4/9

{197} *Stewartia ovata* × 4/9

{198} *Styrax americanus* × 4/9

{199} *Styrax grandifolius* × 1/4

[200] *Leitneria floridana* × ⅗

[201] *Quercus chapmanii* × ⅗

202] *Quercus phellos* × ⅓

[203] *Quercus incana* × ½

[204] *Quercus oglethorpensis* × ¼

[205] *Quercus imbricaria* × ⅜

[206] *Maclura pomifera* × ¼

[207] *Sapium sebiferum* × ⁴⁄₉

[208] *Cotinus obovatus* × ⅐

[209] *Elaeagnus angustifolia* × ⅓

[210] *Lagerstroemia indica* × ½

[211] *Nyssa ogeche* × ²/₇

[212] *Nyssa sylvatica* × ⅕

[213] *Nyssa biflora* × ²/₇

{214} *Cornus alternifolia* × ½

{216} *Diospyros virginiana* × ⅜

{215} *Elliottia racemosa* × ⅕

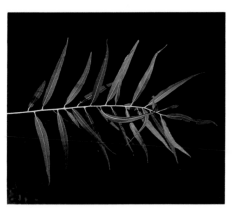

[217] *Salix nigra* × ¼

[218] *Salix caroliniana* × ⅜

[219] *Salix sericea* × ³⁄₁₀

[220] *Salix babylonica* × ⅛

221] *Alnus serrulata* × ²/₇

[222] *Quercus breviloba* × ⅛

[223] *Quercus austrina* × ⅓

[224] *Quercus margaretta* × ½

[225] *Quercus stellata* × ²/₇

[226] *Quercus lyrata* × ¼

[227] *Quercus alba* × ³/₁₀

[228] *Quercus macrocarpa* × ³/₈

[229] *Quercus bicolor* × ¼

[230] *Quercus michauxii* × ⅛

[232] *Quercus prinoides* × ⁴⁄₉

[231] *Quercus montana* × ⅐

[233] *Quercus nigra* × ¼

[235] *Quercus marilandica* × ⅔9

[234] *Quercus arkansana* × ¼

[236] *Quercus ilicifolia* × ½

[237] *Quercus acutissima* × ⅙

[238] *Quercus velutina* × 1⅓

[239] *Quercus falcata* × 2/7

[240] *Quercus pagoda* × ⅓

[241] *Quercus laevis* × ¼

[242] *Quercus rubra* × ⅐

[243] *Quercus georgiana* × ⅖

[244] *Quercus coccinea* × ⅔

[245] *Quercus nuttallii* × 1

[246] *Quercus palustris* × ¼

[247] *Quercus shumardii* × 1⅓

[248] *Pyrus communis* × ⅔

[250] *Malus ioensis* × ⅙

[249] *Malus pumila* × ¼

[251] *Malus angustifolia* × ⅚

[252] *Malus coronaria* × ¼

[253] *Crataegus aestivalis* × ⅖

[254] *Crataegus flabellata* × ⅕

[255] *Crataegus flava* × ⅖

[256] *Crataegus marshallii* × ⅖

[257] *Crataegus spathulata* × ¼

[258] *Crataegus uniflora* × ⅓

[259] *Prunus persica* × ⅕

[260] *Prunus caroliniana* × ³/₇

[261] *Prunus virginiana* × ⅖

[262] *Prunus serotina* × ⅖

[263] *Prunus pensylvanica* × ⅔

[264] *Prunus avium* × 2/7

[265] *Prunus cerasifera* × ¼

[266] *Prunus angustifolia* × ⅜

267] *Prunus maritima* × ²⁄₅

[268] *Prunus mexicana* × ⅕

269] *Prunus americana* × ⅚

[270] *Prunus umbellata* × ²⁄₇

[271] *Ilex decidua* × ⅓

[272] *Ilex verticellata* × ½

[273] *Ilex amelanchier* × ²⁄₉

[274] *Ilex ambigua* × ½

[275] *Ilex montana* × ³⁄₇

[276] *Rhamnus caroliniana* × ⅜

[277] *Franklinia alatamaha* × ¼

[278] *Clethra acuminata* × ²⁄₇

[279] *Oxydendrum arboreum* × ⅕

[280] *Halesia diptera* × ⅖

[281] *Halesia tetraptera* × ⅗

3.
Georgia Hackberry
Celtis tenuifolia Nutt.
[135]

Recognized by leaf blades 2–9 cm long, under 2 × as long as wide, margin entire to sharply toothed except at base, apex acuminate, rarely a few tips falcate; by dried ripe fruits smooth. Shrubs or trees to 12 m tall by 53 cm DBH. Bark smooth when young, later developing corky ridges. Fruits light orange to red or brown, pulp sweet; eaten by birds and various small mammals. Wood rarely used. Occasional plants difficult to separate from *C. occidentalis,* but intergradation does not seem sufficient to treat the two as a single species or varieties of the same species as do some botanists. Common. Dry habitats; rocky soils, bluffs, ridges; in woods or more often in open. Apr–May. Syn: *C. georgiana* Small.

MORACEAE: Mulberry Family

3.
MORUS. Mulberry

Recognized by thornless stems, true terminal buds absent, mature buds with 5–6 exposed scales; by twigs with cloudy or milky juice and continuous pith; by leaves with serrate margin and 3 prominent veins arising from one place at base of blade. Leaves often lobed, especially on vigorous sprouts or young trees. Flowers unisexual, in cylindrical spikes, the male and female flowers on same or different trees. Fruits dense cylindrical clusters of nutlets, each nutlet surrounded by fleshy sepals; popular food of birds and mammals. A genus of about 8 species of trees and shrubs of temperate and subtropical regions of the N. Hemisphere. Two species are native to the US; one of these and an introduced species occur in the SE.

KEY TO MORUS SPECIES
1. Leaves dull green and usually rough above, softly hairy on both large and small veins beneath   1. *M. rubra*
1. Leaves shiny green and smooth above, glabrous beneath except hairs sometimes on main veins and in their axils   2. *M. alba*

1.
Red Mulberry
Morus rubra L.
[136]

Recognized by upper side of leaves dull green and usually rough, the lower side with soft hairs on both large and small veins. Trees to 20 m tall by 1.8 m DBH, with a broad crown. Inner bark fibrous and tough. Largest mature buds 6–8 mm long, somewhat spread-

ing; with 5–6 exposed scales. Leaves often lobed, the lobes one to several. Mature fruits dark dull red to deep purple, sweet and juicy. Wood useful for fence-posts, the heartwood being resistant to decay; other uses such as for furniture and paneling are minor. Trees with unlobed leaves are easily confused with *Tilia,* which may be identified by having only 1–2 exposed scales on mature buds. Common. Mainly in moist soils of lowlands and river valleys and adjacent slopes. Also scattered in various uplands habitats, apparently in most instances having developed from seeds brought in by birds or small mammals. Grows to about 700 m elevation in the sAppalachians. Mar–May.

2.
White Mulberry
Morus alba L.
[137]

Distinguished by leaves shiny, green, and glabrous above, glabrous beneath except sometimes hairs on main veins and in their axils. Trees to 25 m tall by 1.5 m DBH. Largest mature buds 3–4 mm long and ap-pressed. Lobes on leaves none to several. Mature fruits white, pink, or dark purple. Growth form variable; one cultivated form is weeping. White Mulberry was cultivated in China for thousands of years for its leaves, which served as food for silkworms. The species and the silkworms were introduced into the US around the year 1600 in an unsuccessful attempt to establish a silk industry. It has been widely planted and has spread extensively by seeds, in some places a pest because of its abundance and the many juicy fruits dropping in undesirable places, as on sidewalks. Occasional. Around dwellings, fields, pastures, fencerows, vacant lots, second-growth woods. Throughout the SE and beyond. Mostly rare north of NY and in most of wUS except Cal. Mar–May.

LAURACEAE: Laurel Family

4.
SASSAFRAS. Sassafras

Sassafras
Sassafras albidum (Nutt.)
Nees
[138]

Recognized by foliage, buds, and twigs spicy aromatic when crushed or bruised; by leaves more than 2-ranked, lobed or unlobed, with 3 prominent veins aris-ing from same place near base, lacking calluslike growths on upper side in axils of principal veins. Trees to 30 m tall by 1.6 m DBH. First year twigs greenish, smooth or hairy. Leaf blades entire, of 4 shapes: un-

lobed, with two lobes (one on each side), a single lobe on the left side, or a single lobe on the right side. Leaves turning yellow, orange, or red in the fall. Flowers unisexual, the male and female almost always on separate trees. In addition to a pistil, the female flower has 6 sterile stamens. Fruits sit in a red cup on a long stalk; they are blue-black drupes 6–10 mm long, with a shiny brown stone, not abundant, eaten by wild animals. Roots used for tea. Wood soft, little used except for fenceposts, which are durable with bark removed. Common. Thin woods or in open. Fencerows, old fields, clearings in woods, often forming dense stands. Usually to about 1300 m elevation, but here and there to around 1500 m. There are two other species, one in China, one in Taiwan. Mar–May.

HAMAMELIDACEAE: Witch-hazel Family

5.
LIQUIDAMBAR. Sweetgum

Sweetgum
Liquidambar styraciflua L.
[139]

Recognized by leaves star-shaped, with 5 acute nearly equal lobes, occasionally with 2 additional smaller lobes at base, or rarely a few 3-lobed; the margin serrate. Trees to 38 m tall by 1.9 m DBH. Twigs often with corky ridges; end buds true terminal buds; leaf scars with 3 bundle scars. Leaves orange to red or reddish-purple in autumn. Flowers unisexual, on same tree, appearing with the leaves. Male flowers in several clusters on an erect stalk, the female flowers in a somewhat globose cluster at the end of a hanging stalk. Fruits hard ball-shaped structures 2.5–4 cm across, containing many individual seed-bearing sections (capsules); often persistent through winter, as seen in the photograph. Seeds 1–2 in each capsule, a small wing at tip of each seed. Common name derived from a sweet gummy substance produced by the inner bark. Wood with many uses, especially for furniture, plywood, cabinets. Sometimes planted as an ornamental. Common. In woods or open; frequently in old fields and logged areas. Most often in moist soils of bottomlands or floodplains, but occurring in uplands as well. Occurs to about 900 m in sAppalachians. Found in mountains of sMex and Guatemala. The few other species occur in the E. Hemisphere. Feb–May.

FABACEAE: Legume Family

6.
CERCIS. Redbud

Redbud; Judas-tree
Cercis canadensis L.
[140]

Recognized by unlobed heart-shaped leaves with entire margin and 5, sometimes 7, prominent veins arising from one place at base of blade; by petioles swollen at both ends. Shrubs or trees to 14 m tall by 80 cm DBH, with a spreading to rounded crown. Twigs usually zigzag, the end bud axillary, buds with 2 hairy keeled scales, axillary buds often superposed. Flowers normally appearing before the leaves. Fruits flat pods 4–10 cm long, 8–18 mm broad, containing several hard shiny seeds in one row, tardily splitting along one edge, falling in late autumn or winter. Trunks of timber size scarce, the wood rarely used. Planted as an ornamental, but produces many seedlings and can become a pest. Common. Frequent understory tree in hardwood forests, but also occurs in open. Does best in moist habitats, but occurs in dry sites. To about 670 m elevation in SE. Extends into wTex and neMex. There is another species in swUS, 5 in China, and 1 in Europe and wAsia. Feb–May.

EUPHORBIACEAE: Spurge Family

7.
ALEURITES. Tung-tree

Tung-tree
Aleurites fordii Hemsl.
[141]

Recognized by leaves having 5 principal veins arising from one place at base of blade; by a pair of red glands on upper side of the petiole just below the blade. Trees to 10 m tall. Leaf blades broadly ovate, margin entire, generally unlobed on older plants, often with 2 lobes on vigorous shoots and young plants. Flowers unisexual, both sexes on same tree; flowering before or as first leaves develop. Fruits ovoid to globose or obovoid thick-walled capsules 4–8 cm across, containing 3–7 large seeds which are extremely poisonous when eaten. Extensively cultivated locally for the seeds, which yield an oil used in manufacture of paints and varnishes. Rare. Escaped from plantings into various habitats, usually in the open, but also in woods. From s half of La to sGa and nFla. A native of eAsia. Mar–Apr.

8.
MANIHOT. Cassava

Cassava
Manihot grahamii Hook.
[142]

Recognized by leaves lobed nearly to the petiole, the lobes 6–11. Small trees with milky juice, easily killed to the ground by freezing temperatures. Flowers unisexual, both sexes on same tree. Fruits smooth capsules about 15 mm across and on a red disc at end of pedicel. Rare. Escape from cultivation into woods or open, sLa to sGa and nFla. A genus of around 150 species of herbs, shrubs, and trees of the warmer parts of the W. Hemisphere. May–Oct.

RHAMNACEAE: Buckthorn Family

9.
ZIZIPHUS. Jujube

Jujube; Date-tree
Ziziphus jujuba Mill.
[143]

Recognized by leaves glabrous beneath, the margin serrate to crenate, with 3 principal veins arising from base and extending to or near the tip; by a continuous pith. Trees to about 10 m tall by 15 cm DBH, sprouting abundantly from roots, the stems with thorns or rarely unarmed. Leaves glossy above. Flowers small, axillary; petals greenish-white; flowering continuing as fruits develop. Fruits brown to dark-red oblong to ovoid drupes 12–25 mm long, with one stone, the flesh light-colored. Planted for edible fruits and as a curiosity. Rare. Known to have spread by roots and seeds in several localities in Wilkes Co., Ga, and likely elsewhere. Native of eEurope and e and sAsia. May–July.

TILIACEAE: Basswood Family

10.
TILIA. Basswood

American Basswood;
Linden
Tilia americana L.
[144]

Recognized by leaves 2-ranked, unlobed, the margin serrate, with 3–5 principal veins arising at base; by thornless stems, the pith continuous, end buds axillary, exposed bud scales 1–2. Trees to 35 m tall by 2.1 m

DBH. Inner bark tough and fibrous. Leaves variable in shape, especially at the base, and in size, amount, and kinds of hairs. Flowers bisexual, in clusters on a common stalk hanging from a strap-shaped bract. Fruits dry, hard, indehiscent, globose or nearly so, nutlike, usually remaining attached to bract until after it drops, the entire structure falling with a whirling motion that is easily propelled by wind currents, sometimes for considerable distances. Flowers profusely visited by insects, especially honey bees. Attractive as an ornamental and widely planted. Wood especially desirable, including uses for furniture, cabinets, musical instruments. Variation in several features of *Tilia* has led to the naming of many species and varieties, with over a dozen species recognized in each of two older manuals covering the SE. Current data indicate that it is best to recognize them all as a single species until further research resolves this complex problem. *Tilia* is sometimes confused with *Morus,* which is easily distinguished by its 5–6 exposed scales on mature buds. Common. Usually in woods, but also in open; in moist to dry soils. To about 1500 m elevation in sAppalachians. Only species in US; extends into neMex. Probably over 25 species occur in temperate regions of the E. Hemisphere. Apr–July. Some of the more common synonyms are: *T. caroliniana* Mill.; *T. floridana* (V. Engl.) Small; *T. georgiana* Sarg.; *T. glabra* Vent.; *T. heterophylla* Vent.; *T. monticola* Sarg.

MALVACEAE: Mallow Family

11.
HIBISCUS. Hibiscus

Rose-of-Sharon; Althea
Hibiscus syriacus L.
[145]

Recognized by leaves more than 2-ranked, usually lobed, the margin coarsely toothed; by current year's twigs with stellate hairs in lines or sometimes over the entire surface; by flower structure much like that of cotton and hollyhock in that the stamens, usually many, form a column that surrounds but is free from the style of the pistil, the stigmas protruding beyond the column, the column shed with the petals. Shrubs or trees to about 10 m tall. Fruits densely and finely hairy capsules with 5 conspicuous triangular sepals that persist around the base, these subtended by 7–10 linear bracts. There are many horticultural variants based on differences in color and number of petals, shape of leaves, form of plant. Rare. Often planted, sometimes spreading nearby by seeds, occasionally forming thickets. Native of eAsia. May–Sept.

STERCULIACEAE: Sterculia Family

12.
FIRMIANA. Firmiana

**Chinese Parasoltree;
Japanese Varnishtree**
Firmiana simplex (L.)
W. Wight
[146]

Recognized by leaves palmately lobed, with usually 5 principal veins arising from base, margin entire; by buds velvety and reddish-brown. Trees to about 20 m tall by 65 cm DBH. Bark smooth, greenish. Flowers in large terminal clusters; petals absent; sepals strap-shaped, 8–12 mm long, at first yellowish, turning red. The enlarging young fruits split into sections; these in turn split open, appearing like leaves, bearing several seeds on the margin, and becoming leathery at maturity. Planted as an ornamental. Rare. Sometimes escaping near plantings, into woods at some localities; probably not adventive north of La, Tenn, and Va. Escaping in Cal. May–Aug. Syn: *F. platanifolia* (L. f.) Schott & Endl.; *Sterculia p.* L. f.

Group H

Leaves simple, alternate, pinnately veined, evergreen.

12. Vigorous twigs with one leaf at end although leaves often
clustered at tip of short lateral twigs; fruit a berrylike drupe
8. *Ilex*
12. Vigorous twigs with 2–4 leaves clustered at tip; fruit an
acorn   2. *Quercus*
10. Stipules and stipule scars absent   13
13. Largest leaves with few to several conspicuous teeth
16. *Baccharis*
13. Largest leaves entire or, at most, inconspicuously serrate   14
14. Twigs with thorns, sometimes quite scattered; juice milky,
if present   14. *Bumelia*
14. Twigs thornless; any juice not milky   15
15. Netted vein system of leaves conspicuous on upper
surface   16
16. Vigorous twigs with one leaf at end, fruits nearly
dry berries   13. *Vaccinium*
16. Vigorous twigs with 2–4 leaves clustered at tip,
fruits small capsules   7. *Cyrilla*
15. Netted vein system of leaves barely distinct or invisible
on upper surface   17
17. Longest petioles 6 mm or more long; fruits
dehiscent, shallowly 5-lobed capsules   11. *Kalmia*
17. Petioles about 2 mm or less long; fruits indehiscent
3-angled (rarely 2–4) leathery drupes   6. *Cliftonia*

MYRICACEAE: Bayberry Family

1.
MYRICA. Bayberry

Recognized by tiny yellowish to brown glands on underside of leaves. Terminal buds
absent. Flowers small, each subtended by a scale, opening before new leaves appear,
lacking sepals or petals, unisexual, the male and female on separate plants. Male
flowers in dense cylindrical catkins 6–10 mm long and 4–6 mm broad that are
sessile in leaf axils. Female flowers loosely arranged in groups of 2–12 in catkins 5–
10 mm long on peduncles arising from leaf axils. Fruits drupelike, one-seeded,
rounded, with a roughened whitish surface that is coated with wax. This wax can be
removed by hot water, then separated from the water and made into candles. A
genus of about 50 species in temperate and subtropical parts of the world. Two
species attain tree size in the SE: one occurs in the coastal region from Cal to Wash.

KEY TO MYRICA SPECIES

1. Leaves aromatic when crushed, blades flat and usually toothed near end

1. M. cerifera

1. Leaves not aromatic when crushed, blades revolute and entire   2. M. inodora

1.
Southern Bayberry;
Wax-myrtle
Myrica cerifera L.
[147]

Recognized by leaves aromatic when crushed, flat or nearly so, usually toothed near the end. Shrubs or trees to 12 m tall by 32 cm DBH. Leaves to 10 cm long and 2 cm wide, conspicuously reduced in size toward end of most twigs. Part to all of previous year's leaves falling before summer ends. Fruits globose, 2–4 mm across, 1–12 per axillary spike. Often planted as an ornamental, especially in general landscaping. Common. Wet to dry soils, in woods or open. To about 150 m elevation. Occurs in Mex, C. Amer, and W. Indies. Feb–Apr.

2.
Odorless Bayberry
Myrica inodora Bartr.
[148]

Recognized by leaves entire, revolute, and not aromatic when crushed. Shrubs or rarely trees to 7 m tall by 30 cm DBH. Leaves to 10 cm long and 4 cm wide, not conspicuously reduced in size toward end of twigs. Most of previous year's leaves falling before summer ends. Fruits globose, 5–8 mm across, usually one on each axillary stalk. Rare. Swamps, bays, pinelands. Mar–Apr.

FAGACEAE: Beech Family

2.
QUERCUS. Oak

Distinguished by largest of first-year twigs under 3 mm across, pith continuous, stipules and/or stipule scars present, though small; by 2–4 leaves clustered at tip of vigorous twigs, underside lacking scales or glands, foliage not aromatic when crushed. Five species of *Quercus* in the SE are distinctly evergreen. Two others, *Q. incana* and *Q. chapmanii,* sometimes retain leaves through much or all of the winter in the s parts of their range, especially during mild winters, and are included in the following key in case such plants are encountered; but they are generally considered as deciduous and will be described with the other deciduous oaks in Group J. A more detailed description of oaks will be found in Group K.

KEY TO EVERGREEN QUERCUS SPECIES

1. Undersurface of leaves light-colored with tiny dense sessile stellate hairs, these often difficult to see by late summer   2
    2. Hairs on undersurface with rays not parallel with surface, blades with or without a bristle tip (See Group J)   *Q. incana*
    2. Hairs on undersurface with rays parallel with surface except for a small portion of them in No. 2; blades without a bristle tip   3
        3. All hairs on undersurface of leaves stellate and appressed; lateral veins not strongly impressed on upper surface   1. *Q. virginiana*
        3. Hairs on undersurface of leaves stellate and appressed, together with some erect hairs, the latter sometimes sparse in winter; lateral veins strongly impressed on upper surface   2. *Q. geminata*
1. Undersurface of leaves without stellate hairs, either glabrous or with hairs of other types that are usually scattered   4
    4. First-year twigs with a few to many tawny stellate hairs, or closely and finely grayish hairy   5
        5. Hairs on twigs tawny, leaves shiny   3. *Q. myrtifolia*
        5. Hairs on twigs grayish, leaves dull light green (See Group J)   *Q. chapmanii*
    4. First-year twigs glabrous   6
        6. Apex of most leaves obtuse to rounded; upper surface generally dull
                                                                    4. *Q. laurifolia*
        6. Apex of most leaves acute, rarely a few rounded; upper surface usually shiny
                                                                    5. *Q. hemisphaerica*

1.
Live Oak
Quercus virginiana Mill.
[149]

Recognized by undersurface of leaves densely covered with tiny hairs, all appressed and stellate; by lateral veins not strongly impressed on upper surface. Trees to 20 m tall by 3.5 m DBH, with widely spreading, often descending, branches when isolated. Leaves coriaceous, light-colored beneath, commonly entire, unlobed, and the tip rounded, but on vigorous shoots or sprouts may have a sharp tip and marginal teeth; margin sometimes rolled inward, but not hiding any of undersurface. Leaves usually falling just as growth resumes in spring. Acorn cups turbinate, ashy-gray, enclosing ¼ to ½ of the acorn, on stalks 1–13 cm long. Acorns 15–25 mm long; profusely eaten by wild animals. Often planted as an ornamental. Wood dense, but trees suitable for lumber scarce; used mostly for pulp or firewood. Flowering 2–3 weeks before the similar *Q. geminata* when in the same locality and habitat. Common. Usually in dry sandy soils, occasionally in moist habitats, uncommon on dunes. Intolerant of fires. To scarcely 100 m elevation in the SE. Also in neMex. Mar–Apr.

2.
Sand Live Oak
Quercus geminata Small
[150]

Recognized by undersurface of leaves densely covered with tiny appressed stellate hairs and with some erect hairs; by lateral veins strongly impressed on upper surface. Shrubs or trees to about 10 m tall by 60 cm DBH. Leaves coriaceous, light-colored to tannish beneath, entire or rarely slightly lobed, margin revolute and hiding some of the lower surface. Leaves of previous year usually falling shortly after new growth appears in spring. Excellent as firewood; acorns relished by wild animals. Similar to *Q. virginiana* in growth form and acorns but flowering 2–3 weeks later when in same locality and habitat, and hybridization between the two probably minimal. Common. Coastal dunes, inland areas of light-colored sand. Mar–May. Syn: *Q. maritima* (Michx.) Willd.

3.
Myrtle Oak
Quercus myrtifolia Willd.
{151}

Recognized by first-year twigs with a few to many
tawny stellate hairs, often becoming nearly glabrous by
autumn; by leaves glabrous beneath except for tufts of
hairs in axils of main veins. Shrubs or less commonly
trees to 8 m tall by 40 cm DBH, often forming thickets. Leaves shiny, 18–50(80) mm long; most commonly obovate, sometimes ovate to elliptic; usually
unlobed, rarely shallowly lobed, or occasionally undulate; mostly entire and revolute; apex rounded, rarely
obtuse or broadly acute, sometimes bristle-tipped.
Acorn cups sessile or with stalks to 5 mm long, covering ¼ to ⅓ of acorn. Acorns maturing second season,
rarely the first. Of little commercial value; acorns eaten
by wild animals. Similar to *Q. chapmanii,* with which
it is often associated, but which has dull light-green
leaves. Flowering 1–2 weeks after *Q. incana* when in
the same locality. Occasional. Sandy pine or oak-pine
scrub, sandy ridges, the sands usually light-colored.
Mar–Apr.

4.
Diamond-leaf Oak;
Swamp Laurel Oak
Quercus laurifolia Michx.
{152}

Recognized by glabrous first-year twigs; by leaves not
densely hairy beneath, upper surface generally dull,
apex of most leaves obtuse to rounded with a bristle
that usually falls early. Trees to about 45 m tall by 2 m
DBH. A few to most leaves may fall during the winter,
but scattered live leaves are still present when new
spring growth appears. Leaves widest at or near middle, the vein network apparent when observed against a
light source. Leaves of small saplings and stump
sprouts may be coarsely toothed or several-lobed and
the bristles on apex and lobes tend to persist. Acorn
cups deeply saucer-shaped to turbinate, acorns about
15 mm long. Occasionally used as an ornamental.
Wood hard, heavy, and strong, but lumber of only
average quality, used mostly for pulp and a good quality firewood. The acorns a source of food for a variety
of animals. May be confused with *Q. hemisphaerica,* but
that species is separated by having most leaf tips acute
and by flowering about 2 weeks later in same general
area. Common. Bottomlands, poorly drained soils,
swamp margins, rarely in drier soils. Mar–Apr. Syn:
Q. obtusa (Willd.) Ashe.

5.
Laurel Oak
Quercus hemisphaerica
Bartr.
[153]

Recognized by glabrous first-year twigs; by leaves not densely hairy beneath, the upper surface usually shiny; by apex of most leaves acute, rarely some obtuse or rounded. Trees about 40 m tall by 1.5 m DBH. Leaves falling about the time new ones appear in spring, usually widest at or near the middle, the vein network indistinct in transmitted light. Leaves of second growth twigs and of saplings, sprouts, and vigorous shoots commonly with a few coarse teeth or lobes, these and the apex often bristle-tipped. Acorn cups flattened to rounded at base, rarely covering as much as ¼ of the acorn. Widely planted as an ornamental in the CP; wood used mostly for pulp and firewood, the lumber of only average quality; acorns eaten profusely by wildlife. Flowering about 2 weeks later than *Q. laurifolia* in the same general area. Sometimes confused with *Q. phellos,* but in that species the leaf margin is never undulate or lobed, whereas at least some of the leaves of *Q. hemisphaerica* are undulate if not lobed. Common. Sandy dry soils, established dune areas, scrub oak sandhills, stream banks, occasionally in mixed woods. Mar–Apr. Not separated from *Q. laurifolia* Michx. in some treatments.

ILLICIACEAE: Anise-tree Family

3.
ILLICIUM. Anise-tree

Recognized by a strong aromatic odor of all parts when rubbed or bruised; by leaves with lateral veins indistinct or not evident. Shrubs or trees. Leaves crowded at ends of twigs, remaining long after new growth appears in spring, the blades entire and with pale glandular dots beneath the surface. Flowers perfect, solitary on stalks from leaf axils; sepals 3–6, scarcely distinguishable from the 6–30 petals; stamens and pistils numerous, the latter in a single whorl. Fruits dry, star-shaped, with 10–15 sections. There are 7 species, 5 in Asia and 2 in the SE.

KEY TO ILLICIUM SPECIES
1. Leaves acute to acuminate at apex; flowers about 4 cm across, perianth deep red to purplish, petals 20–30   1. *I. floridanum*
1. Leaves obtuse at apex; flowers about 12 mm across, perianth yellow, petals 6–12
                                                                     2. *I. parviflorum*

1.
Florida Anise-tree
Illicium floridanum Ellis
[154]

2.
Yellow Anise-tree
Illicium parviflorum
Michx. ex Vent.
[155]

Recognized by acute to acuminate apex on leaves. Shrubs or trees to 8 m tall by 9 cm DBH. Leaves when crushed or bruised with a strong odor variously described as spicy, fishy, or a mixture of turpentine and spice. Flowers on a downwardly curved stalk from leaf axils, about 4 cm wide, the perianth deep red to purplish or rarely white; petals 20–30, narrow. Fruits 11–15, arranged like the parts of a pinwheel, the entire whorl 25–30 mm across. Seeds flat, somewhat obovoid, very slick, projected several meters as fruit dries. Sometimes planted as an ornamental. Rare. Moist to wet usually wooded habitats; along streams, ravines, swamps, bayheads. Also present in one locality in neMex. Mar–May.

Recognized by leaves with obtuse apex. Shrubs or rarely trees to 10 m tall by 15 cm DBH. Crushed or bruised leaves with odor similar to that of *I. floridanum*. Flowers on downwardly curved stalks from leaf axils, about 12 mm across, the perianth yellow, petals 6–12. Fruits arranged like parts of a pinwheel, the entire whorl 20–25 mm wide. Sometimes planted as an ornamental; a cultivated plant taken from the wild in Liberty Co., Ga, was seen by senior author. Rare and local. Habitats similar to those of *I. floridanum*. May–June.

LAURACEAE: Laurel Family

4.
CINNAMOMUM. Camphor-tree

Camphor-tree
Cinnamomum camphora (L.)
J. Presl
[156]

Recognized by a strong odor of camphor from foliage and stems when crushed or broken; by yellowish callus-like growths in principal vein angles on the upper side of the leaves. Trees to 22 m tall by 3 m DBH, but uncommon over 1 m DBH, the main trunk short. The youngest (outermost) twigs clearly arising from sides of older twigs instead of from the tips as in most tree species. Leaves 5–12 cm long; margin entire, without chlorophyll, somewhat curled so as to appear wavy; midrib wider from its base to the base of the 2 principal lateral veins than beyond these veins. Flowers 2–3 mm long, bisexual, in small stalked axillary clusters; petals greenish-white. Fruits black berrylike globose drupes about 10 mm across, with the taste of camphor; often eaten by birds. Planted as an ornamental. Occasional. Spreading by root suckers from nearby plantings and by seeds, these sometimes dropped by birds in woods and along fencerows at considerable distances. Native of E. Asia. Occurs from extreme sSC to sGa, Fla, and west to Tex; also in sCal. Apr.

5.
PERSEA. Red-bay

Recognized by foliage and twigs with an aromatic odor when crushed or broken; by the lack of glandular dots beneath the leaf surface; by absence of yellowish calluslike growths in principal vein axils on the upper leaf surface. Terminal buds naked; stipule scars absent. Leaf margin entire. Flowers bisexual, in compact clusters on axillary stalks 1–6 cm long. Perianth segments 6, the outer 3 about 2 mm long, the inner 3 about 1 mm long. Stamens 9–12, the inner 3 sterile. Fruits dark blue to black drupes 7–12 mm across, with thin flesh and one stone. Fruits eaten by wildlife, though sparingly; deer browse the foliage. Planted as ornamentals. A genus of about 150 species, mostly of tropical America. Two species are native to the US and both occur in the SE. The Avocado, *P. americana* Mill., has been introduced into subtropical parts of the US and has spread locally from cultivation.

KEY TO PERSEA SPECIES
1. Leaves with hairs lying parallel to undersurface, the hairs either scattered and minute, or dense and clearly visible to the naked eye   1. *P. borbonia*
1. Undersurface of leaves with a few to many hairs that are crooked, variously arranged, occasionally erect, but not appressed   2. *P. palustris*

1.
Red-bay
Persea borbonia (L.)
Spreng.
[157]

Recognized by hairs on underside of leaves straight and parallel to the surface. In var. *borbonia* these hairs are scattered, minute, golden-colored, and barely visible under 10 × magnification, sometimes requiring 40 × to see them clearly. Shrubs or trees to 18 m tall by 60 cm DBH. Common. Coastal dune areas, maritime woods, well-drained sandy soils in woods or open. May–June. The photograph is of this variety. In var. *humilis* (Nash) Kopp, Silk-bay, the hairs on underside of leaves are dense, parallel to the lateral veins, with a silky sheen when young, appearing sooty on old leaves. Shrubs or rarely small trees. Recent evidence indicates that perhaps this variety should be treated as a separate species. Rare. Confined to peninsular Fla, occurring with oak-pine scrub in areas of white sandy topsoil. Apr–May.

2.
Swamp-bay
Persea palustris (Raf.) Sarg.

Similar to Red-bay, but underside of leaves having a few to many hairs that are crooked, variously arranged, sometimes erect, but not appressed, never dense; a magnification of 20 × may be required to see them clearly. Shrubs or trees to 20 m tall by 1 m DBH. Considered by many botanists as a variety of *P. borbonia,* but in our opinion the true relationships are generally misunderstood. Separation by types of hairs is infallible and they almost exclusively occupy different types of habitats. Common. Moist to wet soils, usually in woods; depressions between old stable dunes; less often in drier soils. May–June. Syn: *P. pubescens* Pursh.

CYRILLACEAE: Titi Family

6.
CLIFTONIA. Cliftonia

Titi; Buckwheat-tree
Cliftonia monophylla
(Lam.) Britt. ex Sarg.
[158]

Recognized by thornless twigs with continuous pith, the largest first-year twigs under 3 mm thick, stipules and stipule scars absent; by foliage not aromatic when bruised or crushed; by petioles about 2 mm long or less; by entire-margined leaves with the net-vein system invisible or barely distinct on upper side, the underside glabrous. Shrubs or trees to 9 m tall by about 50 cm DBH, nearly always in dense colonies. Flowers fragrant, in racemes near tip of previous year's twigs, opening in spring before new leaves appear; petals 5, white; stamens 10. Fruits about 7 mm long, strongly 3-angled, uncommonly 2–4 angled, usually some persisting throughout the year until new ones are formed. Grown as an ornamental; nectar an important source for honey. Common. Wet nonalluvial soils, usually in open. Feb–Apr.

7.
CYRILLA. Swamp Cyrilla; Titi

Recognized by thornless twigs with continuous pith, the largest first-year twigs under 3 mm thick, stipules and stipule scars absent, vigorous twigs with 2–4 leaves clustered near tip; by foliage not aromatic when bruised or crushed; by leaves with entire margin, the net-vein system conspicuous on upper surface, the underside glabrous. In colder winters, especially in northern parts of the range, only a few live leaves may remain until new ones appear. Terminal buds glabrous; axillary buds often superposed. Flowers fragrant, bisexual, borne in elongated racemes fastened on or near the apex of the previous year's twigs. Sepals 5, white; petals 5, white; stamens 5, shorter than the petals. Flowers an important source of nectar for honey. Fruits indehiscent, about 2 mm long, with 2 carpels and 2–4 tiny hard seeds, maturing in late summer, often some remaining until flowers appear the next year. A genus of only 2 species, confined to US, W. Indies, C. Amer, and S. Amer to Brazil.

KEY TO CYRILLA SPECIES
1. Leaves mostly 5–10 cm long and 12–25 mm wide, racemes mostly 10–16 cm long, fruits ovoid   1. *C. racemiflora*
1. Leaves mostly 2–4 cm long and 5–12 mm wide, racemes mostly 4–9 cm long, fruits subglobose   2. *C. parvifolia*

1.
Titi; Swamp Cyrilla
Cyrilla racemiflora L.
[159]

Recognized by leaves generally 5–10 cm long and 12–25 mm wide; by racemes mostly 10–16 cm long; by ovoid fruits. Shrubs or trees to 10 m tall by 23 cm DBH. Leaves turning orange to red in late fall. Flowering racemes usually abundant. Tolerant of cold; resistant to pests. Used as an ornamental, doing well in drier upland sites as well as in its usual wet habitat. Common. Stream banks, river swamps, and bottomlands; nonalluvial swamps, pond margins, depressions. May–July.

2.
Little-leaf Cyrilla
Cyrilla parvifolia Raf.

Similar to *C. racemiflora,* but leaves generally 2–4 cm long and 5–12 mm wide, the racemes mostly 4–9 cm long, and fruits nearly globose. Some botanists do not separate these two species; some consider them varieties of the same species. Most populations can be identified by the characters given. In a few colonies plants with intermediate characteristics are common, but such colonies do not seem sufficiently numerous to treat *Cyrilla* as one variable species or 2 varieties of a single species. Rare. Mostly in depressions; sGa and nFla. May–June. Syn: *C. racemiflora* L. var. *parvifolia* Sarg.

AQUIFOLIACEAE: Holly Family

8.
ILEX. Holly; Large Gallberry

Evergreen Species

Evergreen hollies are identified by first-year twigs under 3 mm across, fresh twigs lacking a bitter-almond odor when broken, vigorous twigs with one leaf at end (stubby lateral twigs may have clustered leaves), pith continuous; by tiny stipules that have sharp dark points (these most likely to be observed at junction of current and previous year's twigs, both terminal and lateral ones); by foliage not aromatic when crushed, the underside of leaves lacking scales and yellowish glands. Flowers small, solitary or in clusters from leaf axils, usually of one sex, the sexes on separate plants, or rarely the same. Petals white to greenish-white, 4 or uncommonly 5–6.

Stamens of same number as petals, sometimes present and sterile in the female flowers. Fruits globose or nearly so, berrylike drupes with pulp bitter in most species, with 2–8 bony 1-seeded stones. Evergreen hollies, especially those with red or yellow fruits, used as ornamentals; the fruits an important source of food for wildlife. A genus of over 300 species of shrubs or trees in N. Amer, S. Amer, Asia, and a few in Africa. Six species attaining tree size and occurring in the SE have deciduous leaves and are described in Group K; six species, including one escape, are evergreen.

KEY TO EVERGREEN ILEX SPECIES
1. Leaves with coarse spiny teeth on margin, rarely all leaves entire, at least some leaves with spiny tips   2
   2. Leaves oval to elliptic or elliptic-obovate   1. *I. opaca*
   2. Leaves rectangular-oblong   2. *I. cornuta*
1. Leaves serrate, crenate, or entire, without coarse spiny teeth on margin or tip   3
   3. Leaves crenate along most or all of the margin, sometimes minutely so
                                                                                       3. *I. vomitoria*
   3. Leaves serrate or entire   4
      4. Fruits black, stones smooth on the rounded side; leaves mostly less than twice as long as wide, lateral veins evident on underside   4. *I. coriacea*
      4. Fruits red or yellow, stones with 1 groove or a few ribs on the rounded side; leaves mostly more than twice as long as wide, main lateral veins on underside obscure   5
         5. Leaves 3–10 cm long, 3–30 mm wide   5. *I. cassine*
         5. Leaves 1–4 cm long, 3–7 mm wide   6. *I. myrtifolia*

1.
American Holly
Ilex opaca Ait.
[160]

Recognized by coarse stiff spines on margin and tip of leaves, uncommonly with a spine only on the tip or even some tips spineless; by blades oval to elliptic or elliptic-oblong. Trees to 30 m tall by 80 cm DBH. Flowers about 6 mm across, in stalked clusters from leaf axils, sexes on separate trees, the male in clusters of 3–12, the female solitary or in groups of 2 or 3. Fruits red or orange, rarely yellow, 7–12 mm long, stones 5–6 mm long. Often used as an ornamental, a number of varieties having been developed. Wood whitish, of fine texture, used for such things as carvings and cabinetwork; fruits eaten by wildlife. Common. With mixed hardwoods, occasionally with pines; floodplains to upland sites. To about 1500 m elevation in sAppalachians. Mar–June.

2.
Chinese Holly
Ilex cornuta Lindl.

An introduction which has escaped locally and occasionally reaches tree size. It too has leaves with strong spines, but may be separated by the rectangular-oblong shape. Several horticultural varieties have been developed from this species, Burford being one of the best known. Rare. To be expected as escapes throughout most of the SE. Mar–Apr.

3.
Yaupon; Cassena
Ilex vomitoria Ait.
[161]

Recognized by leaves crenate along most or all of the margin. The crenate aspect may be scarcely detectable on small leaves of plants in especially poor habitats. Shrubs or trees to 14 m tall by 40 cm DBH, sprouting from the roots and tending to form thickets, usually branching near the ground. Leaves leathery, 1–4 cm long, shiny green above. Flowers about 5 mm across, in short clusters from leaf axils, usually unisexual, the sexes on separate trees, the male in dense groups, the female solitary or in groups of 2 or 3. Fruits 4–7 mm across, dull to shiny red, rarely yellow, with 4 narrow stones 3–4 mm long. Often used as ornamentals, as specimen plants, or trimmed into hedges. Wildlife eat the fruits and deer browse the twigs and foliage. Indians, who used the leaves for tea, transplanted the species to new campsites, and some inland populations are probably a result of their introductions. Common. Dune areas, maritime woods, pond margins, swamps, thin upland woods, fencerows. Rarely occurs over 150 m elevation. Mar–May.

4.
Large or Sweet
Gallberry
Ilex coriacea (Pursh)
Chapm.
[162]

Recognized by leaf blades usually less than twice as long as wide, the margin entire or with a few small irregularly and widely spaced teeth above the middle, rarely below; by fruits shiny, black, 7–10 mm across, flesh juicy and sweet, dropping soon after maturing in late summer or autumn, unlike most other *Ilex* species in which fruits persist for a longer period. Shrubs or uncommonly trees to 7 m tall by 10 cm DBH. Young twigs finely hairy, often sticky. Leaves leathery, punctate, the main lateral veins evident on underside. Flowers usually unisexual, the sexes on separate trees, single or in clusters, on stalks 3–9 mm long from axils of leaves, the male flowers few to many in each cluster, the female solitary or 2–3 in a group. Fruits an important source of food for wildlife. Common, but tree size rare. Flatwoods, depressions, bays, stream swamps, pond margins. Mar–May.

5.
Dahoon
Ilex cassine L.
[163]

Recognized by stiff leaves 3–10 cm long, 3–31 mm wide, the blades mostly more than twice as long as wide, margin entire or occasionally with one to few irregularly and widely spaced small teeth above the middle; by fruits 5–7 mm across, bright red, sometimes orange, uncommonly yellow. Shrubs or trees to 22 m tall by 25 cm DBH. Flowers about 5 mm across, solitary or in stalked clusters, usually unisexual, the sexes on separate trees; male flowers many per cluster; female flowers solitary or in groups of 2–3. Fruits 5–7 mm across, usually abundant, some persisting until spring, the pulp mealy and bitter; eaten by wildlife. Occasional. Wet soils in woods; depressions, marsh margins, stream banks, cypress ponds, swamps, bays; also with mixed hardwoods. In scattered localities in Cuba and ceMex. Apr–May.

6.
Myrtle-leaved Holly
Ilex myrtifolia Walt.
[164]

Recognized by rigid leaves 1–4 cm long, 3–7 mm wide, the blades mostly twice as long as wide; margin revolute, entire or occasionally with a few small sharp teeth toward apex; by fruits 5–8 mm across, red, rarely orange or yellow. Shrubs or less commonly trees to 14 m tall by 50 cm DBH. Flowers on short stalks, unisexual, sexes on separate trees. Fruits often persisting to spring, pulp mealy and bitter; eaten by wildlife. Occasionally used as an ornamental. Occasional. Cypress and gum ponds or depressions, margins of sandy ponds, moist sandy pinelands. Apr–June.

THEACEAE: Tea Family

9.
GORDONIA. Gordonia

Loblolly-bay; Red-bay
Gordonia lasianthus (L.)
Ellis
[165]

Recognized by first-year twigs 3 mm or more across; by leaves with small blunt teeth on the margin, not aromatic when crushed. Trees to 25 m tall by 1.2 m DBH. Crown narrow, compact. Twigs glabrous; termi-

nal and axillary buds without scales, with shiny silky hairs. Flowers bisexual, 5–10 cm across, solitary on stalks 3–7 cm long. Petals with silky hairs on back; stamens numerous, the filaments united at base into a shallow 5-lobed cup. Fruits ovoid, 1–2 cm across, splitting into 5 sections at maturity. Seeds several, brown to black, shiny, winged. Deer browse on the foliage. Not easily grown under cultivation. Loblolly-bay is related to *Franklinia,* the "lost" *Gordonia.* Common. Wet habitats; flatwoods, depressions, bays, cypress and gum swamps, pond margins, low broadleaf woods. Occurs to about 150 m elevation. There are about 15 other species, which occur in seAsia. May–Oct.

ERICACEAE: Heath Family

10.

RHODODENDRON. Rhododendron

Recognized by first-year twigs 3 mm or more across; by leaves with entire margin, not aromatic when crushed. Flower buds terminal, conical, 25–37 mm long; leaf buds smaller. Corollas bell-shaped, with five lobes; stamens 10, of different lengths. Fruits 5-celled capsules, splitting into 5 sections at maturity, releasing the many tiny flattened seeds. Deer browse the leaves, which are poisonous when eaten in large amounts. Wood used in making bowls of briar pipes; a few lesser uses include tool handles and crafts. There are over 600 species of rhododendrons, evergreen or deciduous, shrubs or trees, occurring in the cool and temperate regions of the N. Hemisphere and one extending into nAustralia. None of the deciduous species reaches tree size in the US; 2 evergreen species reach tree size in the SE; another occurs in the Pacific Coast region from cwCal into BC.

KEY TO RHODODENDRON SPECIES

1. Most or all leaf blades under 90° at base; inner bark of current year's twigs usually reddish   1. *R. maximum*
1. Most or all leaf blades over 90° at base; inner bark of current year's twigs usually greenish   2. *R. catawbiense*

I.
Rosebay
Rhododendron;
Great-laurel
Rhododendron maximum L.
[166]

Recognized by most or all leaf blades acute, under 90°, at base; by inner bark of current year's twigs usually reddish. Shrubs or trees to about 10 m tall by 30 cm DBH. In subfreezing temperatures or during severe droughts leaf blades roll into a coil and bend downward. Flowers about 4 cm across, in terminal clusters of 10–30; calyx lobes mostly 2–5 mm long; the corolla white to pale-pink or rarely reddish-pink, with greenish-yellow spots on upper lobe, pedicels and ovaries with stalked glands. Fruits oblong, 9–13 mm long, widest below middle, with sticky hairs. Often used as an ornamental. Common. Stream banks, ravines, slopes, swamps, most often in hardwood forests. Mostly below 1000 m, but occasionally to over 1800 m in sAppalachians. May–Aug, some scattered flowers as late as early Oct.

2.
Mountain Rosebay;
Purple-laurel
Rhododendron catawbiense
Michx.
[167]

Recognized by most or all leaf blades obtuse, over 90°, at base; by inner bark of current year's twigs usually greenish. Shrubs or uncommonly trees to 7 m tall by 10 cm DBH. Leaves tend to roll and droop under freezing conditions or during severe droughts. Flowers about 6 cm across, in terminal clusters; calyx lobes 0.5–1 mm long; corolla deep pink to purple, rarely nearly white; pedicels and ovaries finely hairy to glabrous. Fruits 10–18 mm long, oblong, widest below the middle, not glandular. Widely used as an ornamental along with a considerable number of hybrids with other, mostly exotic, species. Common. In hemlock, northern hardwood, and spruce-fir forests; heath balds or "slicks" on mountain ridges. Usually in the mountains above 500 m, reaching 1980 m, but in scattered localities as low as 60 m. Apr–June, to July at highest elevations, with a few flowers until early Oct.

II.
KALMIA. Kalmia

Mountain-laurel
Kalmia latifolia L.
[168]

Recognized by largest first-year twigs under 3 mm across, pith continuous, stipules and stipule scars lacking, branchlets thornless; by foliage not notably aromatic when crushed; by longest petioles 6 mm or more long; by leaves with entire margin, the net-vein system on upper surface invisible or barely distinct, undersur-

face lacking scales or glands. Shrubs or trees to 9 m tall by 30 cm DBH, the trunk crooked. Twigs sticky and hairy when young, later smooth. Leaves often clustered near end of twigs; occasional leaves on a plant may be opposite or in 3's, but most are alternate. Flowers in large terminal clusters or in smaller clusters from the axils of leaves of the previous year, corolla 2–3 cm across, white to deep pink. The 10 stamens are bent so that the anthers fit into small pockets in the corolla; when touched by an object, such as an insect, the stamen springs toward the stigma, releasing pollen on the stigma and/or object. Fruits broader than long, shallowly 5-lobed, splitting open into 5 sections in the late fall; some likely to remain until the following summer. Leaves poisonous to livestock when eaten in liberal amounts, but deer browse on them with little effect unless the leaves are a large portion of the diet. Often used as an ornamental. Wood used for briar tobacco pipes and in craft items. Common. An understory plant or a component of open heath balds. In well-drained upland ravines, bluffs, or sandy stream terraces in the SE; may occur in bogs and near sea level in ne parts of its range. To over 1200 m in the sAppalachians. Mar–July.

12.
LYONIA. Lyonia

Tree Lyonia;
Staggerbush
Lyonia ferruginea (Walt.)
Nutt.
[169]

Recognized by undersurface of leaves with grayish to rusty scurfy scales; by foliage and twigs not notabiy aromatic when crushed or broken. Shrubs or trees to 12 m tall by 23 cm DBH with crooked trunk. Leaves stiff, ruffled or wavy, 2.5–9 cm long, 1–5 cm wide, margin entire and revolute, apex acute to obtuse and mucronate, all of about equal size from base to tip of twig. Flowers borne in clusters from axils of previous year's leaves, corolla 2–4 mm long, stamens 10. Fruits dry, ovoid to ellipsoid, 3–6 mm long, 5-carpelled, splitting open at maturity, some fruits persisting until after flowers appear the following year. Tree Lyonia is suspected of being poisonous when eaten, but its foliage is not very palatable. In its shrubby form easily mistaken for the shrub *L. fruticosa* (Michx.) Torr. ex B. L. Robins., but latter may be separated by flowers usually on present year's growth, occasionally on twigs of the previous year, leaves conspicuously reduced in size toward ends of twigs, and flowering May–July. *Bumelia tenax,* with its rust-colored leaf undersurface, may be confused with Tree Lyonia, but this appearance is due to hairs and not scales. Common. There are about 35 species of *Lyonia* occurring in Asia, US, Mex, and W. Indies; 5 species occur in US, all east of the Rocky Mts., but only 1 reaches tree size. Well-drained soils of oak forests, pine-oak scrub, dry hammocks, dunes, raised sandy areas in flatwoods. Mar–Apr, uncommonly to May.

13.
VACCINIUM. Blueberry

Sparkleberry;
Farkleberry
Vaccinium arboreum Marsh.
[170]

A thornless plant recognized by largest first-year twigs under 3 mm across, vigorous ones with 1 leaf at tip, stipules and stipule scars absent; by foliage not notably aromatic when crushed; by net-vein system on upper leaf surface conspicuous, the underside glabrous or with fine hairs on the veins, margin entire or with tiny teeth. Shrubs or trees to 10 m tall by 35 cm DBH with crooked branches. Terminal buds absent. Leaves 2–7 cm long, usually shiny above, sometimes dull and glaucous. Flowers in racemes that also bear green bracts or small leaves, corolla bell-shaped, 5–8 mm across and about as long, stamens 10, ovary inferior. Not all flowers open at the same time on a given plant and begin opening earlier on some plants than others. Fruits shiny black berries 5–8 mm long, with scanty slightly sweet mealy pulp and 8–10 seeds; ripening in autumn, but persisting into winter. Wildlife eat the berries, deer browse the foliage; the wood hard and tough, used locally for tool handles and craft items. Common. Usually in dry sandy and rocky areas in open or in woods; occasionally in moist situations. To 790 m elevation in sAppalachians. *Vaccinium* is a large widely distributed genus of around 150 species, occurring from high mountains of the tropics to north of the Arctic Circle. Only Sparkleberry reaches tree size in the US. Mar–July.

SAPOTACEAE: Sapodilla Family

14.
BUMELIA. Bumelia

Recognized by twigs with thorns, these sometimes quite scattered and occasionally some bearing leaves; by juice milky and sticky, pith continuous, largest first-year twigs under 3 mm across, stipules and stipule scars absent; by foliage not aromatic when bruised, leaf margin entire. Terminal buds absent; axillary buds very small, embedded in bark; thorns lateral to buds. Leaves often clustered on short lateral spurs. Flowers bisexual, in clusters of 8–40 from axils of leaves or leaf scars, most often closely grouped on spurs which may have few to rarely no leaves. Petals united, the lobes about as long as the tube. Stamens 10, short, 5 fertile and 5 sterile, all fastened to the corolla. Ovary superior. Fruits drupelike 1-seeded black berries with thin pulp. A genus of about 30 species from Brazil and W. Indies to the US; 4 species attain tree size in the US; all 4 occur in the SE, 3 within the range of this book, the fourth in sFla, sTex, and Mex.

KEY TO BUMELIA SPECIES
1. Leaves glabrous beneath or only slightly hairy   1. *B. lycioides*
1. Leaves quite hairy beneath   2
   2. Hairs on lower surface of leaf shiny, closely appressed, with most hairs lying nearly parallel with the lateral veins   2. *B. tenax*
   2. Hairs on lower surface of leaf woolly, not shiny, and not all closely appressed, tending to spread in different directions laterally   3. *B. lanuginosa*

1.
Buckthorn Bumelia
Bumelia lycioides (L.) Pers.
[171]

Recognized by leaves glabrous beneath or nearly so. Shrubs or trees to about 15 m tall by 18 cm DBH. Twigs tough, flexible; winter buds to 3 mm long. Leaves elliptic or oblanceolate, 5–10 cm long, with net-vein system prominent. Some of the leaves normally shed over winter; severe winters in northern range limits can result in much shedding with only a few live leaves persisting. Corollas 3–5 mm long. Fruits ellipsoid to ovoid or obovoid, 8–20 mm long, with a large nearly globose hard seed; eaten by wildlife. Rare. Plants scattered, rarely in groups; usually in understory. Swamps, stream borders, lake margins; less often on slopes, bluffs, dunes. To about 300 m elevation. Apr–July.

2.
Tough Bumelia
Bumelia tenax (L.) Willd.
[172]

Separated from other bumelias by presence on lower leaf surface of dense closely appressed hairs, most of them lying nearly parallel with the lateral veins; hairs shiny and silvery, pale gold, copper, or rusty-brown; in dune areas the hairs may be dull-colored and badly worn due to blowing sand. Shrubs or trees to 12 m tall by 25 cm DBH with tough flexible twigs. Leaves 3–7 cm long, mostly oblanceolate, the tip rounded. Fruits dull, 10–14 mm long; eaten by wildlife. Occasional. Dunes, maritime woods, sandy pinelands, scrub oak sandhills. May–June.

3.
Gum Bumelia
Bumelia lanuginosa
(Michx.) Pers.

Having leaves similar to those of *B. tenax,* but the dense hairs on the undersurface woolly, dull, and tending to spread in different directions laterally. Shrubs or trees to 25 m tall by 65 cm DBH. Common. Sandy woods, dry rocky soils; also borders of swamps and streams. Occurs westward to seAriz and neMex. Apr–July.

SYMPLOCACEAE: Sweetleaf Family

15.
SYMPLOCOS. Sweetleaf

Sweetleaf; Horse-sugar
Symplocos tinctoria (L.)
L'Her.
[173]

Recognized by pith of twigs chambered; by foliage not notably aromatic when bruised, leaves finely hairy beneath. Shrubs or trees to 17 m tall by 36 cm DBH. Largest first-year twigs under 3 mm across, terminal buds with acute tip, scales ciliate. Leaves 7–15 cm long, margin entire or occasionally some teeth on the

apical half, with a sweet taste that may be faint in old leaves. Conspicuous when in flower; flowers opening before new leaves develop, fragrant, in clusters from axils of previous year's leaves or from just above the leaf scars if the leaves have fallen; petals creamy yellow to yellow, pistil 1. Fruits nearly cylindrical to ellipsoid drupes 8–12 mm long, with thin pulp and a hard stone containing 1 seed; the tip usually retaining parts of the sepals. Foliage relished by browsing wildlife. A yellow dye may be obtained from bark and leaves. Occasional. Plants often scattered, uncommonly grouped. Thin to dense woods of slopes, bluffs; broad-leaf woods of sandy soils, stream borders; stable dunes. A genus of about 380 species of the warm parts of Asia, Australia, the Americas; only this species occurs in N. Amer, mostly in the SE. Mar–May.

ASTERACEAE: Composite Family

16.
BACCHARIS. Baccharis

Eastern Baccharis;
Silverling
Baccharis halimifolia L.
[174]

Recognized by first-year twigs under 3 mm across, stipules and stipule scars absent; by foliage not notably aromatic when bruised; by largest leaves with a few to several conspicuous teeth. Shrubs or trees to 9 m tall by 16 cm DBH. Twigs ribbed, long remaining green. Foliage olive-green in winter; those leaves near flowers often entire. Flowers tiny, in dense heads, unisexual, the sexes on separate plants. The plant in the photograph has male flowers; female flowers are more conspicuous. Fine bristles associated with female flowers 10–15 mm long, white or nearly so, becoming quite conspicuous as the tiny dry 1-seeded fruits at the bristle bases mature. Poisonous when eaten. Common. On raised areas in and by salt marshes, thin woods, stream margins, bay shores, fencerows, abandoned fields. Presumably from generally level parts of the CP near or not far from the sea, but now widely spread inland and continuing to spread. Perhaps as many as 300 widely distributed species, but this is the only one attaining tree size in the US. Also occurs in the Bahamas and Cuba. Aug–Nov.

Group I

Leaves simple, alternate, deciduous, 2-ranked.

KEY TO GENERA

1. Leaves entire, rarely otherwise; buds without overlapping scales   2
  2. Leaf bases asymmetrical, lateral buds stalked   9. *Hamamelis*
  2. Leaf bases symmetrical, lateral buds sessile   3
    3. Crushed or bruised leaves ill-scented; end bud a terminal one   8. *Asimina*
    3. Crushed or bruised leaves not ill-scented; end bud an axillary one
        12. *Styrax*
1. Leaves serrate or crenate to dentate; buds with or without overlapping scales   4
  4. Lateral buds superposed   12. *Styrax*
  4. Lateral buds solitary, not superposed   5
    5. Lateral buds stalked; leaf margin wavy to coarsely crenate to crenate-dentate
        9. *Hamamelis*
    5. Lateral buds sessile; leaf margin serrate to dentate   6
      6. Teeth of leaf margin of same number as main lateral veins, a lateral vein ending at tip of each tooth   4. *Castanea*
      6. Teeth of leaf margin more abundant than main lateral veins, at least some teeth lacking a main lateral vein   7
        7. Lenticels on vigorous 2–3-year-old twigs lengthened horizontally, distinct   3. *Betula*
        7. Lenticels on vigorous 2–3-year-old twigs circular, often indistinct or not evident   8
        8. End bud a terminal one   9
          9. Bundle scars 3; visible bud scales several, overlapping, hairs sometimes present but not long silky ones   10. *Amelanchier*
          9. Bundle scar 1; visible bud scales usually 2, opposing each other, with long silky hairs   11. *Stewartia*
        8. End bud an axillary one   10
          10. Leaves symmetrical at base   11
            11. Trunk with flaking bark, not fluted; buds not 4-angled, scales striated   2. *Ostrya*
            11. Trunk smooth and fluted with musclelike ridges; buds 4-angled, scales not striated   1. *Carpinus*
          10. Leaves asymmetrical at base (*Planera* may have most leaves symmetrical, but not all)   12
            12. Widest part of leaf about midway or beyond   5. *Ulmus*
            12. Widest part of leaf below middle, usually about ¼ of length from base   13
              13. Tips of leaves acute, rarely some short-acuminate
                6. *Planera*
              13. Tips of leaves acuminate   7. *Zelkova*

BETULACEAE: Birch Family

I.
CARPINUS. Hornbeam

American Hornbeam;
Blue-beech
Carpinus caroliniana Walt.
[175]

Recognized by bark smooth and with rounded irregularly spaced vertical ridges; by lenticels on 2−3-year-
old twigs circular, if present; by end bud an axillary
one; by buds sessile, 4-angled, only 1 in each leaf axil;
by leaf blades symmetrical at base, the margin doubly
serrate. Trees to 20 m tall by 70 cm DBH, the trunk
often crooked. Twigs slender, slightly zigzag; axillary
buds sharp-pointed, scales not striated; stipules present. Flowers tiny, unisexual, the sexes on same tree;
male flowers in dense drooping cylindrical catkins 3−4
cm long, appearing in the spring; female flowers fewer,
arranged in shorter and less conspicuous catkins. Fruits
4−6 mm long, ovoid, paired in axil of a bract that has
a long terminal lobe and much smaller lateral lobes
near the base. Wood close-grained and very hard, tree
sometime called "Ironwood"; used locally for tool handles. Similar to *Ostrya,* which may be separated by its
flaky bark and by leaves with several of the lower major
lateral veins branched. Common. Stream banks and
other moist rich soils, usually as an understory tree. To
about 900 m elevation in sAppalachians. A genus of
about 25 species of Europe and Asia, this the only one
native to N. Amer. In addition to the eUS, it occurs
from cMex to Honduras. Feb−Apr.

2.
OSTRYA. Hophornbeam

Eastern Hophornbeam
Ostrya virginiana (Mill.)
K. Koch
[176]

Recognized by bark with narrow shreds; by lenticels on
vigorous 2−3-year-old twigs circular, if present; by end
bud an axillary one; by buds sessile, not 4-angled, only
1 in each leaf axil; by leaf blades symmetrical at base,
the margin doubly serrate. Trees to 22 m tall by 90 cm
DBH. Axillary buds pointed, scales striated; stipules
present. Dead brownish leaves often persist on trees
during winter. Flowers tiny, unisexual, sexes on same
tree; male flowers in drooping narrow cylindrical
catkins visible in winter, in spring enlarging to 4−6
cm long as flowers open; female flowers in slim cylindrical catkins about 2 cm long or less. Fruits small
ovoid ribbed nuts 5−6 mm long enclosed in thin inflated sacs, these arranged in elongated hanging clusters 3−5 cm long, having the appearance of hops,

which suggests its name. Flowers and fruits eaten by
birds. Wood quite hard, tree often called "Ironwood";
used locally for tool handles. Similar to *Carpinus,*
which may be separated by its smooth bark and few, if
any, branches on the lower major lateral veins of the
leaves. Common. Rich low woods to well-drained
slopes and ridges; usually under taller broadleaf trees.
To 1370 m elevation in sAppalachians. A genus of
about 8 species occurring in Europe, Asia, and N.
Amer; 2 in wUS. This species extends to nNS and west
to neWyo; in scattered localities in Mex and south to
Honduras. Dec–May.

3. BETULA. Birch

Recognized by lenticels on vigorous 2–3-year-old twigs elongated horizontally; by
lateral buds 1 in each leaf axil; by teeth on leaf margin more abundant than the
main lateral veins. Small to large trees. End bud on vigorous twigs axillary; pith
triangular. Lowermost lateral veins of leaves of some species, especially *B. populifolia,*
so prominent that they might be placed in Group G. Because of this, *Betula* is
included in the key to genera of that group, but all species are described under
Group I. Flowers unisexual, the sexes on the same tree. Male flowers in slender
elongated partially developed catkins that appear in the fall, enlarging as flowers
open in the spring. Female flowers also in slender catkins, but smaller and developing on short new twigs in the spring. Fruits small, with 2 lateral wings, borne in
dense cones, each fruit in the axil of a bract. A genus of about 50 species of trees
and shrubs occurring in the N. Hemisphere from temperate to subarctic regions.
Fourteen species are native to N. Amer from Tex to the Arctic Circle; 7 of these are
trees, of which 6 occur in the SE and most extend beyond. An introduced species,
B. alba L., has become naturalized locally in the SE.

KEY TO BETULA SPECIES
1. Leaf blades suborbicular or nearly so, apex rounded (very rare, native of swVa)
                                                                    3. *B. uber*
1. Leaf blades of other shapes, apex acute to acuminate   2
    2. Crushed twigs, especially the bark, with odor and flavor of wintergreen;
        fruiting catkins sessile or nearly so   3
        3. Buds broadly acute; bark yellowish to silvery-gray, separating in thin layers;
            scales of fruiting catkins 6–13 mm long, finely hairy, margin ciliate, 3-
            lobed with lateral lobes ascending   1. *B. alleghaniensis*
        3. Buds sharply acute; bark reddish-brown, tight, resembling that of *Prunus
            serotina,* furrowed on the larger trunks; scales of fruiting catkins 5–7 mm
            long, glabrous, 3-lobed with lateral lobes divergent   2. *B. lenta*

2. Crushed twigs without wintergreen odor or flavor, usually bitter; fruiting catkins peduncled   4

   4. Leaves ending in prolonged taillike tip, glabrous beneath; bark of young stems remaining tight   5. *B. populifolia*

   4. Tip of leaves acute to acuminate, glabrous or hairy beneath; bark of young stems exfoliating   5

      5. Buds shiny with resin, leaf apex acute (rare escape)   6. *B. alba*

      5. Buds not resinous, leaf apex acute to acuminate   6

         6. Leaves usually with 5–8 pairs of lateral veins, base of blades on each side of petiole diverging from lowest lateral vein; bark on young trees chalky to creamy white; fruiting catkins drooping or spreading

                                            4. *B. papyrifera*

         6. Leaves usually with 8–12 pairs of lateral veins, base of blades on each side of petiole entire and parallel to lowest lateral vein; bark on young trees pinkish to tan or reddish-brown; fruiting catkins erect or nearly so   7. *B. nigra*

1.
Yellow Birch
Betula alleghaniensis Britt.
[177]

Recognized by bark yellowish to silvery-gray, separating in thin layers; by crushed twigs, especially the bark, with odor and taste of wintergreen; by mature buds broadly acute; by leaves acute to short-acuminate at apex; by scales of fruiting catkins 6–13 mm long, finely hairy, the margin ciliate, 3-lobed with lateral lobes ascending. Trees to 35 m tall by 1.4 m DBH, often occurring in dense stands. Leaves a rich yellow-gold in autumn. Seeds sometimes germinating on fallen logs which later rot away leaving the birches with trunks on aboveground prop-roots. An important source of lumber with many uses including cabinet work, interior finishes, veneers, and tool handles. Flowers, buds, and fruits eaten by birds. Young plants lacking the distinctive bark and fruiting cones may be mistaken for Sweet Birch, but its bud scales are glabrous while those of Yellow Birch have fine hairs and are ciliate. Common. Rich woods of mountain slopes and ravines; in spruce-fir forests. To 1950 m elevation in sAppalachians. Extends to se and swNfld. Apr–May. Syn: *B. lutea* Michx. f.

2.
Sweet Birch;
Cherry Birch
Betula lenta L.
[178]

Recognized by bark reddish-brown and tight, resembling that of *Prunus serotina,* furrowed on the larger trunks; by crushed twigs, and especially the bark, with odor and taste of wintergreen; by mature buds sharply acute; by leaves acute to short acuminate at apex; by scales of fruiting catkins 5—7 mm long, glabrous, 3-lobed, the lateral lobes divergent. Trees to 27 m tall by 1.5 m DBH, usually solitary. Leaves a rich golden yellow in autumn. The lumber valuable, with uses similar to those of Yellow Birch, though not as abundant. Once the source of oil of wintergreen, which is now manufactured almost exclusively from chemicals. Birds eat the buds, flowers, and fruits; browsing animals utilize twigs and leaves. Young plants lacking the distinctive bark and fruiting cones may be mistaken for Yellow Birch, which see for distinguishing characters. Common. Rich woods, but sometimes in open rocky areas, such as balds. To 1830 m in sAppalachians. Mar—May.

3.
Ashe's Birch
Betula uber (Ashe) Fern.

Easily recognized by its nearly circular leaves with rounded tips. Rare. In understory of mixed broadleaf trees. Only a very few trees exist, these in Smyth Co., swVa.

4.
Paper Birch
Betula papyrifera Marsh.
[179]

Recognized by bark of young trees chalky to creamy white, separating into papery strips, on larger trunks becoming dark and often furrowed; by crushed twigs lacking wintergreen odor and taste; by buds not resinous; by leaf tips acute to acuminate. Trees to 24 m tall by 67 cm DBH. Leaves ovate to nearly triangular, margin doubly serrate. Male flowers in narrow dense drooping catkins 7—10 cm long, the female flowers in much smaller erect catkins. Fruits in narrow catkins 2—5 cm long, hanging on a slender stalk. Widespread and quite variable, with perhaps 6 varieties worth recognition throughout its range. The SE plants are var. *cordifolia* (Regel) Fern. in NC, with cordate leaf bases and var. *papyrifera* in WVa, with leaf bases cuneate to rounded. Of little value in SE because of scarcity. Rare. Usually in coniferous forests, from about 1500—2000 m in NC. Extends to Nfld and Alas, south to neOre. May—July.

5.
Gray Birch
Betula populifolia Marsh.

Similar to *B. papyrifera,* having smooth chalky to
grayish-white bark, but without papery strips. The
leaves differ in having a prolonged taillike tip. Trees to
17 m tall by 37 cm DBH. Of little commercial value
in SE. Rare. Usually in dry poor soils, abandoned
fields, thin woods. Grows to about 600 m elevation.
Extends to nNS. June–July.

6.
European White Birch
Betula alba L.

A native of Europe and Asia that has been planted as
an ornamental and escaped locally. It may be recognized by white peeling bark, buds shiny with resin,
and leaves with an acute tip. Rare. June.

7.
River Birch
Betula nigra L.
[180]

Recognized by bark pinkish to tan or reddish-brown,
separating into thin papery scales, on older trunks becoming dark-brown and with coarse scales; by crushed
twigs lacking odor of wintergreen; by leaves with acute
tips, base of blade on each side of petiole entire and
parallel to lowest lateral vein. Trees to 29 m tall by 1.3
m DBH. Male flowers in dense slender hanging catkins
7–10 cm long; female flowers in smaller upright
catkins; mature fruiting catkins brownish, upright,
2.5–4 cm long, releasing tiny 2-winged fruits in late
spring or early summer. Wood of poor quality for
lumber due to numerous knots from the many
branches; has some value as pulpwood. Common.
Lowlands, floodplains, and banks of streams and rivers;
pond and lake margins; occasionally scattered in upland sites. To about 750 m elevation in sAppalachians
to near sea level in neUS. Feb–Apr.

FAGACEAE: Beech Family

4.
CASTANEA. Chestnut; Chinquapin

Recognized by teeth on leaf margin of same number as main lateral veins, a lateral
vein ending at tip of each tooth, the teeth usually bristle-tipped; by lateral buds
sessile, with 2–3 visible bud scales, the end bud an axillary one. Shrubs or trees.
Pith of twigs 5-angled. Flowers unisexual, both sexes on same plant. Male flowers
strong-scented, tiny, usually closely arranged on long narrow spreading to upright
catkins. Female flowers on short catkins or along the lower part of male catkins.
Fruits are nuts, 1–5, surrounded by a tough spiny husk that splits into 2 or 4
sections at maturity. A genus of about 12 species in temperate regions of Asia,
nAfrica, sEurope, and eN. Amer, 3 species being native to the SE; 1 introduced
species has escaped locally from plantings.

KEY TO CASTANEA SPECIES
1. Underside of leaves with dense tiny velvety hairs   2
    2. Leaves acute to obtuse or rounded at apex, rarely acuminate; fruiting burs to 4
       cm across, usually with 1 nut   1. *C. pumila*
    2. Leaves acuminate at apex, sometimes shortly so; fruiting burs 4–7 cm across,
       usually with 2–3 nuts   2. *C. mollissima*
1. Underside of leaves nearly glabrous or, if obviously hairy, not densely or velvety

                                                                                        3
    3. Leaves without stellate hairs beneath; bur splitting into 4 sections, with 3
       (rarely 2–5) nuts   3. *C. dentata*
    3. Leaves with at least a few stellate hairs beneath; bur splitting into 2 sections,
       with 1 nut (rarely 2)   4
        4. Leaves acute to obtuse or rounded at apex, rarely acuminate; twigs tan to
           yellow-green, slender   1. *C. pumila*
        4. Leaves mostly acuminate at apex; twigs dark brown to gray-brown, stout
                                                                      4. *C. ozarkensis*

1.
Chinkapin; Chinquapin
Castanea pumila (L.) Mill.
[181]

Recognized by leaves acute to obtuse or rounded at
apex, with scarce to abundant tiny stellate hairs beneath; by fruiting burs to 4 cm across, splitting into 2
sections, with 1 nut (rarely 2). Shrubs or trees to 15 m
tall by 80 cm DBH. Twigs usually finely hairy. Burs
with hairy spines; nuts to 19 mm broad. The photograph shows persistent upper portions of the narrow
catkins with remains of male flowers still present. Nuts
food for people and wildlife; little use made of the
wood, except perhaps for fenceposts. Plants resistant to
the blight that has almost wiped out the formerly very
common American Chestnut. A variable species with

intergrading forms some of which, in the past, have been treated as separate varieties or species. The form in the photograph, with the body of the bur readily visible between the spines, occurs in the CP and has been named var. *ashei* Sudw. Plants farther inland have the body of the bur hidden by the spines; this form has been treated as var. *pumila*. Other forms have been given names, including *C. alnifolia* Nutt. and *C. floridana* (Sarg.) Ashe. Common. Thin woods or in open; dry rocky, sandy, or loamy soils; well-drained stream terraces, dry pinelands, sandhills. To about 1350 m elevation; apparently absent from some sections of the sAppalachians. Apr–July.

2.
Chinese Chestnut
Castanea mollissima Blume
[182]

Recognized by leaves with tiny dense velvety hairs on undersurface, the apex acuminate, sometimes shortly so. Trees to about 25 m tall. Leaves 8–18 cm long. Burs 4–7 cm across, usually with 2–3 nuts. Little affected by chestnut blight. Native of eAsia and planted for its nuts. Rare. Naturalized near plantings at scattered localities. June.

3.
American Chestnut
Castanea dentata (Marsh.)
Borkh.
[183]

Recognized by leaf blades usually over 15 cm long, the apex acuminate, undersurface, if hairy, the hairs not stellate; by burs splitting into 4 sections, with 3 (rarely 2–5) nuts. Trees once attained heights of over 35 m and DBH over 3 m, but few plants now reach tree size due to destruction by chestnut blight that was accidentally introduced into New York City in 1904. Bur body hidden by glabrous spines; however, burs now uncommon. Wood once the main source of tannin; quite resistant to decay and a few split-rail fences and perhaps some railroad ties still exist. The nuts once an important source of food for man and wildlife. Rare as a tree, occasional as smaller plants, usually stump sprouts and seldom over 5 m tall. With other broadleaf trees and pines on mountain slopes, hillsides, and other well-drained sites. June–July.

4.
Ozark Chinquapin
Castanea ozarkensis Ashe

Also having leaves with blades usually over 15 cm long and an acuminate apex. It may be recognized by the bur body being clearly visible through the hairy spines. Shrubs or trees formerly to 20 m tall by 1.2 m DBH. Leaves with at least a few stellate hairs beneath, sometimes glaucous. Burs splitting into 2 sections, with 1 nut (rarely 2). Highly susceptible to chestnut blight, now mostly stump sprouts. Occasional. Dry slopes, rocky ridges, well-drained stream terraces. June–July. Syn: *C. alabamensis* Ashe; *C. arkansana* Ashe.

ULMACEAE: Elm Family

5.
ULMUS. Elm

Recognized by each end bud an axillary one; by lenticels on vigorous 2–3-year-old twigs circular to vertical short oblong; by leaves asymmetrical at base or barely so in 2 introduced species, widest about midway or beyond, margin serrate to crenate or doubly serrate, the teeth more abundant than lateral veins which are straight and parallel. Buds oblique, directed to one side, bud scales 4–8 in 2 ranks. Stipules present. Flowers small, bisexual, in small clusters in spring from buds on twigs of previous year or in late summer from buds in axil of current year's leaves. Fruits flat, 1-seeded, with a thin wing encircling the seed cavity; eaten by wildlife. Deer browse on leaves and twigs. Wood hard, tough, and difficult to split, thus useful for tool handles, various implements, and crating; also used for pulp and paneling. There are over 25 species of *Ulmus,* all restricted to temperate parts of the N. Hemisphere. Six species are native to N. Amer, all are restricted to eN. Amer and all occur in the SE. Three other species, exotics, have escaped from plantings in the SE.

KEY TO ULMUS SPECIES
1. Leaves simply serrate or those on vigorous twigs sometimes partly doubly serrate, less than half the primary teeth with a small secondary tooth; fruits glabrous throughout   2
   2. Bark nearly smooth, shedding a few thin scales; buds about 2 mm long; flowering and fruiting in late summer   1. *U. parvifolia*
   2. Bark rough and furrowed; buds to 4 mm long; flowering and fruiting in spring   2. *U. pumila*
1. Leaves doubly serrate; fruits glabrous or hairy on margins and/or body   3
   3. Buds with many, usually appressed, reddish-brown hairs, at least on

uppermost scales; leaves very rough hairy above; inner bark mucilaginous, not
bitter 3. *U. rubra*
3. Buds glabrous or with small grayish to brownish hairs; leaves smooth to rough
 hairy above; inner bark not mucilaginous, usually bitter 4
 4. Branches without corky wings, the bark uniform around the branches; body
 of fruit glabrous 5
 5. Leaves mostly over 8 cm long; flowers hanging from slender stalks to 2
 cm long; margin of fruits ciliate, but the body glabrous
 4. *U. americana*
 5. Leaves mostly under 8 cm long; flowers in short-stalked clusters; margin
 of fruits glabrous 5. *U. procera*
 4. Some or all branches with narrow to broad corky wings, these sometimes
 scarce or rarely absent on slow-growing and old trees; body of fruits hairy,
 sometimes finely so 6
 6. Flowers and fruits produced in the spring 7
 7. Vegetative buds about 2 mm long, acute but not sharply pointed,
 nearly twice as long as broad, scales not ciliate; largest leaves 4−8 cm
 long, to 11 cm on vigorous stems 6. *U. alata*
 7. Largest vegetative buds 3−6 mm long, ovoid, sharply pointed, less
 than twice as long as broad, scales ciliate; largest leaves 8−14 cm
 long 7. *U. thomasii*
 6. Flowers and fruits produced in summer or autumn 8
 8. Leaves 5−9 cm long, smooth above; margin of fruits with hairs about
 1.0 mm long 8. *U. serotina*
 8. Leaves mostly under 5 cm long, rough above; margin of fruits with
 hairs about 0.5 mm long 9. *U. crassifolia*

1.
Chinese Elm
Ulmus parvifolia Jacq.
[184]

Recognized by nearly smooth bark that sheds a few
thin scales; by leaves simply serrate or, on vigorous
twigs, less than half of the teeth with a small secondary
tooth; by flowering and fruiting in late summer, the
fruits glabrous throughout. Trees to 18 m tall by 60
cm DBH. Buds about 2 mm long. Some leaves persisting into winter in warmer parts of SE. A native of
China, introduced as an ornamental. Resistant to dutch
elm disease. Rare. Escaping locally from plantings
throughout most of the SE. Scarce to absent in mountains. Aug−Sept.

2.
Siberian Elm
Umus pumila L.

Similar to *U. parvifolia* in that leaves are simply serrate
and the fruits glabrous throughout, but may be recognized by having rough and furrowed bark and flowering and fruiting in spring. Trees to 35 m tall by 1.2 m
DBH. A native of Asia used sparingly as an ornamental. Resistant to dutch elm disease. Rare. Escaping
from plantings in scattered localities in the SE. Mar−
Apr.

3.
Slippery Elm; Red Elm
Ulmus rubra Muhl.
[185]

Recognized by buds with many, usually appressed, reddish-brown hairs, at least on the uppermost scales; by inner bark mucilaginous, not bitter; by leaves doubly serrate. Trees to 27 m tall by 1.9 m DBH, usually scattered, less commonly in colonies. Bark deeply and irregularly furrowed. Leaves quite rough above. Fruits nearly circular, 12–20 mm long, slightly notched at tip, finely hairy over seed cavity, otherwise glabrous. Somewhat susceptible to dutch elm disease. Common. Usually in lowlands or on rich wooded hillsides and in ravines, occasionally in dry upland sites. Uncommon above 600 m elevation. Jan–Apr. Syn: *U. fulva* Michx.

4.
American Elm;
White Elm
Ulmus americana L.
[186]

Recognized by bud scales glabrous or with a few light-colored hairs; by branches without corky wings; by leaves mostly over 8 cm long, doubly serrate; by margin of fruits ciliate, the body glabrous. Trees to 28 m tall by 2.5 m DBH. Trunk often divided into 2 equal forks and these in turn divided. Flowers and fruits hanging from slender stalks to 2 cm long. Once abundant along city streets, in parks, and as a shade tree, but dutch elm disease, introduced accidentally about 1930, and another disease, phloem necrosis, have devastated the species. Millions of dollars have been spent annually removing dead trees. Occasional, once common. Bottomlands, swamp forests, floodplains, ravines, rich woodlands. To about 750 m elevation. Extends to nNS and to seSask and extreme neWyo. Jan–Apr. Syn: *U. floridana* Chapm.

5.
English Elm
Ulmus procera Salisb.

Similar to American Elm in lacking corky wings on branches, having doubly serrate leaves, and body of fruits glabrous. It may be separated by bud scales minutely pale hairy; by leaves mostly under 8 cm long; by flowers and fruits in short-stalked clusters, margin of fruits glabrous. Rare. Introduced as an ornamental and escaped locally. Apr.

6.
Winged Elm
Ulmus alata Michx.
[187]

Recognized by some or all branches with corky wings, these sometimes scarce on slow-growing and old trees; by vegetative buds about 2 mm long, acute, but not sharp-pointed, the scales not ciliate, flower buds larger; by leaves doubly serrate, the largest 4–8 cm long, to 11 cm on sprouts and vigorous stems; by flowers and fruits produced in spring. Trees to 35 m tall by 1 m DBH. Leaves smooth to quite rough above. Fruits 6–10 mm long, the margin with cilia over 1 mm long, the sides finely hairy, apex deeply notched. Common. In woods or in open; usually in uplands, but occasionally in moist habitats; fencerows, old fields. Rarely over 600 m elevation. Jan–Mar.

7.
Rock Elm; Red Elm
Ulmus thomasii Sarg.
[188]

Recognized by branches with corky wings, these sometimes absent; by buds 3–6 mm long, sharply pointed, the scales ciliate; by leaves doubly serrate, the largest 8–14 cm long; by flowers and fruits produced in the spring, in clusters to 4 cm long. Trees to 30 m tall by over 1 m DBH. Leaves smooth above, with fine soft hairs beneath. Fruits 10–19 mm long, the margin with hairs under 1 mm long, finely hairy on the sides, the apex finely notched. Common. In woods or open; ridges, hillsides, flatlands, limestone outcrops. To about 750 m elevation. Apr.

8.
September Elm;
Red Elm
Ulmus serotina Sarg.
[189]

Recognized by branches with corky wings, these sometimes quite scattered; by leaves 5–9 cm long, margin doubly serrate, smooth above; by flowers and fruits produced in summer or autumn, sometimes after leaves have fallen; by hairs on margin of fruits about 1 mm long. Trees to 21 m tall by 80 cm DBH. Twigs with scattered whitish lenticels, buds about 5 mm long. Fruits elliptic, 10–14 mm long, apex deeply notched. Highly susceptible to dutch elm disease. Once used occasionally as a street tree, in parks, and around residences. Rare. Limestone areas, bottomlands, hillsides. Sept–Oct.

9.
Cedar Elm
Ulmus crassifolia Nutt.

Having characters similar to those of September Elm including corky wings on twigs and flowers and fruits developing in autumn. Cedar Elm may be distinguished by leaves mostly under 5 cm long, rough above; by hairs on margin of fruits about 0.5 mm long. Trees to 28 m tall by 80 cm DBH. Common. Along streams, bottomlands, upland sites, frequently in limestone areas. To about 460 m elevation. Extends into neMex. Sept–Oct.

6.
PLANERA. Water-elm

Water-elm; Planertree
Planera aquatica
J. F. Gmel.
[190]

Recognized by leaf blades with widest part distinctly below the middle, margin serrate to doubly serrate, base asymmetrical on some to all leaves, the apex acute. Trees to 23 m tall by 80 cm DBH, usually with a short trunk, often with sprouts from or near base. Bark thin, gray to light brown, with age developing conspicuous flat scales which fall exposing the reddish-brown inner bark. Branches slender; terminal bud absent. Some flowers male, with 4–5 stamens, others on the same tree female, or more frequently bisexual with a pistil and a stamen. Fruits 1-seeded nutlike drupes bearing irregularly shaped projections; commonly eaten by waterfowl. Occasional. Usually in areas subject to temporary flooding; swamps, wet banks of rivers and creeks, floodplains. This is the only species of the genus. Feb–Apr.

7.
ZELKOVA. Zelkova

Japanese Zelkova
Zelkova serrata (Thunb.)
Makino
[191]

Recognized by leaves with widest part below the middle, margin serrate, base asymmetrical in some or all leaves, the apex acuminate. Trees to 23 m tall by 60 cm DBH. Bark gray, smooth, developing thin scales with age. Crown dense, broad, and rounded. Leaves rough above. Flowers unisexual, the sexes on same tree, the male in axils near stem base, the female in leaf axils nearest the stem tip. Fruits oblique-shaped drupes, nearly sessile. A native of Japan; planted as an ornamental. More resistant to insect damage than elms. Rare. Local escape from plantings. A genus of 5 species of Asia and eEurope. May.

ANNONACEAE: Custard-apple Family

8.
ASIMINA. Pawpaw

Recognized by buds with reddish hairs and no scales, the lateral buds sessile and often superposed, end bud a terminal bud; by crushed foliage and stems ill-scented; by leaves entire, the base symmetrical. Shrubs or trees. Twigs with continuous and diaphragmed pith, sometimes becoming chambered. Axillary buds tilted. Flowers bisexual, with 3 sepals, 6 petals in 2 unequal-sized whorls of 3 each; stamens many, spirally arranged; pistils 3–15. Fruits irregularly cylinder-shaped berries with 1–several large hard smooth brown seeds. Two species reach tree size in the SE.

KEY TO ASIMINA SPECIES
1. Flowers under 2 cm broad, leaves 6–20 cm long   1. *A parviflora*
1. Flowers 2–5 cm broad, leaves 15–30 cm long   2. *A triloba*

1.
Dwarf Pawpaw
Asimina parviflora
(Michx.) Dunal
[192]

Recognized by flowers less than 2 cm broad, pedicels less than 1 cm long at time of pollen shedding. Shrubs or trees to 6 m tall by 14 cm DBH. Buds reddish-hairy. Leaves 6–20 cm long, upper surface glabrous, the lower surface densely reddish-hairy when young to sparsely red-hairy on veins with aging. Flowers often appearing before leaves. The photograph shows 2 flowers, one with all parts present (not all visible), another with sepals, petals, and stamens fallen, the several growing pistils present. Fruits 3–7 cm long, greenish-yellow when ripe, with dark brown seeds 1–2 cm long; eaten by wildlife. Occasional. Rich woods, margins of lime sinks, sandy or dry woods, hammocks. Feb–May.

2.
Pawpaw
Asimina triloba (L.) Dunal
[193]

Recognized by flowers 2–5 cm broad, pedicels 1 cm or more long at time of pollen shedding. Shrubs or trees to 16 m tall by 88 cm DBH. Buds with dark brown hairs. The photograph shows in particular the distinctive buds, which lack scales. Leaves 15–30 cm long, upper surface with a few appressed reddish hairs when young, the lower surface densely so; becoming glabrous above and sparsely hairy on veins beneath. Flowers maroon, appearing before leaves. Fruits oblong-cylindric, 5–15 cm long, yellow-green to brownish when ripe; with brown to dark brown, bean-shaped seeds 1.5–2.5 cm long, shiny when mature. Fruits eaten by wildlife. Occasional. Rich hardwood forests, riverbottoms. To 790 m in sAppalachians. Feb–May.

HAMAMELIDACEAE: Witch-hazel Family

9.
HAMAMELIS. Witch-hazel

Witch-hazel
Hamamelis virginiana L.
[194]

Identified by lateral buds stalked, naked, solitary in leaf axils; by leaves oblique at base, margin entire or wavy to coarsely crenate to crenate-dentate. Shrubs or trees to 13 m tall by 30 cm DBH. Trunk usually crooked, often more than one, frequently branching near the ground. Twigs zigzag; buds hairy, terminal ones 6–13 mm long. Leaves 6–15 cm long. Flowers bisexual, solitary or in groups of 2–3 per leaf axil, often appearing after leaves drop in autumn. Fruits densely hairy capsules 10–15 mm long with 4 sharp curved points on the end, maturing about Oct–Nov, splitting into 2 parts and ejecting the blackish seeds with considerable force. An aromatic oil extracted from leaves, twigs, and bark used on a limited basis in lotions and certain patent medicines. A few species of wildlife eat the seeds or the entire fruits; deer browse the foliage. Some botanists recognize *H. vernalis* Sarg. as another species of Witch-hazel. It flowers Jan–Apr, has calyx lobes reddish to orange on the inner surface, petals are mostly less than 12 mm long, and almost all plants are shrubs. *H. virginiana* flowers Sept–Dec, has calyx lobes greenish to yellowish on the inner surface, petals are usually over 12 mm long, and it is more often a tree. The former was thought to occur only west of the Mississippi R. and to be generally distinct from the latter but more recently plants with the characteristics of *H. vernalis* have been reported from La and sAla to swFla and sGa. In addition, many of the plants have borne characteristics of both species so that maintaining *H. vernalis* as separate is now questionable. Common. Rich to dry woods; ravines, rocky slopes, sandy ridges, steep slopes, sandy pinelands, stream banks. Extends to NS; a small population occurs in neMex. To about 1800 m in sAppalachians but uncommon above 1500 m. There are 4 other species in eAsia. Sept–Dec, occasionally Jan–Apr in s and w parts of range.

ROSACEAE: Rose Family

10.
AMELANCHIER. Serviceberry; Shadbush

Recognized by lateral buds solitary, sessile, somewhat curved, with several visible overlapping scales; by bundle scars 3; by lenticels on 2–3-year-old twigs circular, if evident; by leaves serrate, the teeth abundant. Shrubs or trees. Terminal buds acuminate; stipules present but soon dropping. Flowers perfect, in terminal racemes, appearing before or with the leaves. Fruits pomes 6–10 mm across; the flesh usually soft, juicy, and sweet, ripening in early summer; eaten profusely by birds. Seeds several. Identification is often difficult and the number of species in doubt, perhaps 25 native to N. Amer. Serviceberries occur north to Alas and south to Guatemala; there are 4 species in Eurasia; 2 species of tree size occur in the SE.

KEY TO AMELANCHIER SPECIES
1. Leaves ovate to elliptic, pointed at apex, teeth of largest leaves 5–12 per cm; tip of ovary glabrous   1. *A. arborea*
1. Leaves usually elliptic to nearly circular, usually blunt to rounded at apex, teeth of largest leaves 4–6 per cm; tip of ovary hairy   2. *A. sanguinea*

1.
Serviceberry; Shadbush
Amelanchier arborea
(Michx.f.) Fern.
[195]

Recognized by leaves ovate to elliptic, pointed at apex, teeth of largest leaves 5–12 per cm. Shrubs or trees to 17 m tall by 80 cm DBH. Leaves 4–10 cm long. Petals 10–17 mm long, strap-shaped, white to tinted with pink. There are 2 varieties. In var. *arborea* [*A. canadensis* (L.) Medic.] mature leaves hairy beneath becoming nearly glabrous with age, the apex usually acute; flowering racemes upright. Most abundant at low elevations in the mountains and southward. In var. *laevis* (Wieg.) Ahles leaves glabrous except when young, the apex usually acuminate, rarely acute; racemes divergent to drooping. Most abundant in the mountains to higher elevations and northward. Sometimes planted as an ornamental; wood of little value. Common. Rocky woods, mountain balds, river banks, wooded slopes, swamp margins. To 1960 m elevation in sAppalachians. Extends to sNfld. Feb–May.

2.
Roundleaf Serviceberry
Amelanchier sanguinea
(Pursh) DC.

Separated from the above species by having elliptic to circular leaves with blunt to rounded tips, teeth of largest leaves 4–6 per cm. Shrubs or trees to 11 m tall by 13 cm DBH. Rare. Rocky slopes. To 610 m elevation in SE. Apr–June.

THEACEAE: Tea Family

11.
STEWARTIA. Stewartia

Recognized by lenticels on 2–3-year-old twigs circular, if evident; by lateral buds sessile and solitary in leaf axils; by visible bud scales usually 2, opposing each other, with long silky hairs; by finely serrate leaves. Evergreen or deciduous shrubs or trees. Bundle scar one, raised; stipules none. Flowers bisexual, occurring singly in

leaf axils. Stamens numerous, falling with the 5 petals; pistil 1. Fruits hard capsules, splitting into 5 sections, releasing the several seeds. There are about 10
species, 8 native to seAsia and 2 deciduous species native to the SE. Flowers of the 2
SE species are showy and flowering plants resemble *Cornus florida* from a distance.

KEY TO STEWARTIA SPECIES

1. Pistils with 5 united styles; seeds shiny; fruits with rounded lobes, very short-
   pointed   1. *S. malacodendron*
1. Pistils with 5 separate styles; seeds dull; fruits strongly angled, prominently
   pointed   2. *S. ovata*

1. **Virginia Stewartia; Silky-camellia** *Stewartia malacodendron* L. [196] 	Recognized by pistils with 5 united styles; by fruits with rounded lobes, short-pointed. Shrubs or trees to 6 m tall by 10 cm DBH. Petioles stout, 4–9 mm long. Flowers about 10 cm broad, on stalks less than 1 cm long, petals with wavy edges, stamens purple. Seeds shiny. Planted as an ornamental, but slow-growing. Rare. Usually under other broadleaf trees; wet to dry soils, often sandy. To about 150 m elevation. Apr–June.
2. **Mountain Stewartia** *Stewartia ovata* (Cav.) Weatherby [197] 	Recognized by pistils with 5 separate styles; by fruits strongly angled, prominently pointed. Shrubs or trees to 7 m tall by 7.5 cm DBH. Petioles usually over 7 mm long. Flowers about 10 cm broad, stalks 1–2 cm long, petals with wavy margins, stamens yellow, or purple in an uncommon form. Seeds dull-colored. Occasionally planted as an ornamental. Occasional. Usually under other broadleaf trees and in rich soils, sometimes in moist soils; steep slopes, bluffs. To about 750 m elevation in sAppalachians. June–July. Syn: *S. pentagyna* L'Her.

STYRACACEAE: Storax Family

12.
STYRAX. Storax; Snowbell

Recognized by buds naked, sessile, scurfy; by each end bud an axillary one; by leaves entire, the base symmetrical, not ill-scented when crushed or bruised. Twigs with minute stellate hairs; 1–3 buds in leaf axils, the buds superposed. Flowers fragrant; the calyx 5-lobed; petals white, united at base, with 5 narrow lobes; stamens 10, united near their bases. Fruits capsules, obovoid to nearly globose, 5–9 mm across, usually 1-seeded, ordinarily dropping with seed enclosed. About 100 species of shrubs or trees of tropical and warm climates, most species in Eurasia, 4 in the W. Hemisphere. Two species reach tree size in the SE.

KEY TO STYRAX SPECIES
1. Leaves elliptic to oval, obovate, or lanceolate, 1–4 cm wide; flowers solitary or in racemes of 2–4   1. *S. americanus*
1. Leaves broadly oval to broadly obovate or nearly orbicular, 3–9 cm wide; flowers 5–20 in racemes that are sometimes leafy   2. *S. grandifolius*

1.
Storax; American Snowbell
Styrax americanus Lam.
[198]

Recognized by leaves 1–4 cm wide; by flowers solitary or in racemes of 2–4. Shrubs or trees to 5 m tall by 9 cm DBH. Leaves elliptic to oval, obovate, or lanceolate, nearly glabrous beneath, with a few to several small stellate hairs, mostly in the veins. Corolla lobes 8–12 mm long, with small hairs. Fruits to about 7 mm across, densely short-hairy. Common, but rare as a tree. Moist to wet places; bottomland woods, floodplains, swamps, stream banks. Mostly under 600 feet elevation. Feb–June. Syn: *S. pulverulenta* Michx.

2.
Bigleaf Snowbell
Styrax grandifolius Ait.
[199]

Recognized by leaves 3–9 cm wide; by flowers 5–20 in racemes which are sometimes leafy. Shrubs or trees to 8 m tall by 10 cm DBH. Leaves broadly oval to broadly obovate or nearly orbicular, with abundant small stellate hairs, these sometimes becoming few with age. Corolla lobes 10–20 mm long, finely hairy. Fruits to about 12 mm across, densely short-hairy. Foliage quite similar to that of *Halesia diptera,* the two being difficult to distinguish without flowers or fruits. Both have superposed buds, but these differ in position and shape. In *H. diptera* the buds are contiguous and the top one is triangular in outline. In *S. grandifolius* there is a definite space between the upper two buds and the top one is thumblike in shape. Occasional, but rare as a tree. Deciduous or mixed woods, usually in well-drained areas, rarely in moist to wet places. To 1000 m elevation. Mar–May.

Group J

Leaves simple, alternate, deciduous, more than 2-ranked, entire, unlobed.

KEY TO GENERA

1. Leaf blades as wide as long or nearly so, the tips acuminate; a pair of glands on upper side of juncture of petiole and blade   5. *Sapium*
1. Leaf blades considerably longer than wide or, if about as wide as long, the tips obtuse to rounded; glands lacking at juncture of petiole and blade   2
   2. First-year twigs and lower surface of leaves covered with scurfy scales   7. *Elaeagnus*
   2. First-year twigs and leaves without scurfy scales   3
     3. End bud a true terminal one   4
       4. Fresh foliage, buds, and twigs spicy-aromatic when crushed or tasted (See Group G)   *Sassafras*
       4. Fresh foliage, buds, and twigs not spicy-aromatic   5
         5. Fresh foliage and twigs with bitter-almond taste and odor when crushed or broken (See Group K)   *Prunus*
         5. Fresh foliage and twigs lacking such a taste or odor   6
           6. Pith chambered, foliage sweet to taste (see Group H)   *Symplocos*
           6. Pith continuous, foliage not sweet   7
             7. Pith with diaphragms, bundle scars 3   9. *Nyssa*
             7. Pith without diaphragms, bundle scars 1 to many   8
               8. Buds with 2 nearly valvate scales, internodes clearly unequal   10. *Cornus*
               8. Buds with several scales, internodes of most genera approximately equal   9
                 9. Underside of leaves velvety to touch; bundle scars 3 and distinct; leaf tips acute   1. *Leitneria*
                 9. Underside of leaves glabrous or, if velvety hairy to touch, bundle scars several and indistinct; leaf tips acute to rounded   10
                   10. Terminal and larger lateral buds and their scales rounded to acute, but none acuminate   2. *Quercus*
                   10. Terminal and larger lateral buds and their scales acuminate and sharp pointed   11
                     11. Leaves obtuse to rounded at apex, bundle scars more than 1   6. *Cotinus*
                     11. Leaves acute, bundle scar 1   11. *Elliottia*
     3. End bud an axillary one   12
       12. Stipules and stipule scars absent   13

13. Pith diaphragmed, sometimes partly chambered; fruits conspicuously fleshy berries, to 4 cm across, with 1 to several large flat seeds   12. *Diospyros*
13. Pith not diaphragmed; fruits berries with scanty flesh, to 2 cm across   14
   14. Twigs with thorns, sometimes quite scattered, sap milky; leaves dull above; fruits drupelike 1-seeded berries (See Group H)   *Bumelia*
   14. Twigs without thorns, sap not milky; leaves glossy above; fruits several-seeded berries (See Group H)   *Vaccinium*
12. Stipules and stipule scars present   15
   15. First-year twigs winged or angled, at least just below each leaf; fruits nearly globose capsules   8. *Lagerstroemia*
   15. First-year twigs not winged or angled; fruits fleshy   16
      16. Bundle scar 1; twigs without milky sap or thorns; fruits berrylike drupes, 3–10 mm across (See Group K)   *Ilex*
      16. Bundle scars several; twigs with milky sap and thorns (sometimes sparse); fruits multi-sectioned, 2–15 cm across   17
         17. Leaves not lobed, fruits 7–15 cm long   3. *Maclura*
         17. Leaves sometimes 3-lobed toward end, fruits to 3.5 cm long (rare escape)   4. *Cudrania*

LEITNERIACEAE: Corkwood Family

1.
LEITNERIA. Corkwood

Corkwood
Leitneria floridana Chapm.
[200]

Recognized by twigs with pith continuous and without diaphragms, lacking scurfy scales and bitter-almond taste and odor, with 3 depressed bundle scars in leaf scar, the buds hairy and with several scales; by leaves considerably longer than wide, the underside velvety hairy. Shrubs or trees to 7 m tall by 12 cm DBH, usually in thickets. Bark thin; branches few; terminal buds conic, about 4 mm long. Leaves thick and leathery when mature, the veins depressed in upper surface. Flowers unisexual, the sexes on separate plants, both sexes borne in ascending dense scaly cylindrical catkins. Fruits drupes, a little flattened, ellipsoid to oblong, 16–25 mm long, 1-seeded. Wood very light in weight; has been used for fish-net floats and bottle stoppers. Rare. Wet places, swamps, marshes, river and stream banks, saline shores. Feb–Mar.

FAGACEAE: Beech Family

2.
QUERCUS. Oak

Species with Deciduous Entire Leaves

These species are distinguished by twigs with pith continuous and without diaphragms, bundle scars in leaf scars several and usually indistinct; by buds with
several overlapping scales, the larger buds rounded to acute at apex; by 2–4 leaves
clustered at tip of vigorous twigs; by leaf blades considerably longer than wide,
lacking scurfy scales, with no bitter-almond odor when crushed. Acorns of all species
are an important source of food for wildlife. There are 10 species of oaks occurring in
the SE that have at least some individual plants with deciduous entire leaves and all
are included in the accompanying key. Five of these are sufficiently consistent in
their characteristics to be described in Group J. Two species are normally evergreen
and are described in Group H, but atypical circumstances may result in their
appearing deciduous. Three species, *Q. nigra, Q. sinuata,* and *Q. breviloba,* have leaf
forms that might easily be mistaken for Group J; however, they more generally fit
Group K and are described there. A more detailed description of oaks is found in
Group K.

KEY TO QUERCUS SPECIES
1. Most or all leaves widest toward tip   2
    2. Leaves wedge-shaped at base (See Group K)   *Q. nigra*
    2. Leaves blunt to rounded at base   1. *Q. chapmanii*
1. Most or all leaves widest at or near middle   3
    3. Leaves glabrous beneath at maturity, except perhaps tufts of hairs in axils of
       lateral veins   4
        4. Apex of most or all leaves obtuse to rounded   5
            5. Bark of trunk dark, tight, ridged and furrowed on larger trunks; mature
               acorns on second-year twigs (See Group H)   *Q. laurifolia*
            5. Bark of trunk light-colored, with thin loosely appressed scales; mature
               acorns on twigs of current year   6
                6. Acorn cups flattened at base, covering less than ¼ of acorn (See
                   Group K)   *Q. sinuata*
                6. Acorn cups rounded at base, covering ¼ to ⅓ of acorn (See Group K)
                                                                            Q. breviloba
        4. Apex of most or all leaves acute, rarely some obtuse or rounded   7
            7. Leaves thin and pliable, dull above, net vein system on upper surface
               usually not prominently raised   2. *Q. phellos*
            7. Leaves thick and coriaceous, shiny above, net vein system on upper
               surface prominently raised (see Group H)   *Q. hemisphaerica*

3. Leaves hairy on lower surface, but often appearing glabrous to naked eye, especially late in year 8
    8. Hairs on underside of leaves stellate and sessile 9
        9. Rays of stellate hairs not parallel to leaf surface 3. *Q. incana*
        9. Rays of stellate hairs parallel to leaf surface (See Group K) *Q. breviloba*
    8. Hairs on underside of leaves stellate, the rays on a common stalk 10
        10. Bark gray; leaves without a bristle tip 4. *Q. oglethorpensis*
        10. Bark brownish; leaves with a bristle tip, these often dropping with age 5. *Q. imbricaria*

1.
Chapman Oak
Quercus chapmanii Sarg.
[201]

Recognized by leaves widest toward the tip, usually obovate, base obtuse or more commonly rounded to somewhat cordate. Some leaves oblong to elliptic and unlobed, more commonly irregularly wavy-margined to shallowly lobed toward the apex. Shrubs or trees to 18 m tall by 40 cm DBH. First-year twigs closely and finely grayish hairy, the cuticle rarely visible beneath the hairs. Cups covering ⅓ to ½ of acorn, acorns 15–25 mm long. Occasional. Sandy barrens and scrub, dunes, ridges. Feb–Mar.

2.
Willow Oak
Quercus phellos L.
[202]

Recognized by leaves widest near the middle, thin and pliable, the apex of most or all acute, the net vein system on upper surface usually not prominently raised. Trees to 34 m tall by 2.3 m DBH. First-year twigs slender, terminal buds sharp-pointed. Leaves 5–8 × as long as wide with no lobes or undulations, glabrous above, the undersurface glabrous or hairy in vein axils or along midrib, turning pale yellow in fall. Acorn cups covering ¼ to ⅓ of acorn, scales hairy; acorns finely hairy, maturing second year. Wood used for lumber, pulp, and firewood. Often planted as a shade tree, especially along streets, around public buildings, and in parks. Common. Bottomlands, floodplains and adjacent slopes, rich uplands. To 1000 m elevation. Mar–May.

3.
Bluejack Oak; Upland
Willow Oak
Quercus incana Bartr.
[203]

4.
Oglethorpe Oak
Quercus oglethorpensis
Duncan
[204]

5.
Shingle Oak
Quercus imbricaria Michx.
[205]

Recognized by most or all leaves widest at or near the middle, the underside bearing grayish sessile stellate hairs with the rays not parallel to the surface. Shrubs or trees to 15 m tall by 65 cm DBH, sometimes forming dense thickets. Twigs with dense minute hairs; terminal buds narrowly lanceolate to narrowly ovate, pointed, scales hairy. Leaves mostly narrowly elliptic to elliptic, 5–11.5 cm long, rarely with 1–4 short bristle-tipped lobes. Acorn cups 10–18 mm broad, sessile or nearly so, covering ¼ to ⅓ of acorn, scales appressed. Acorns finely hairy, at least toward summit, maturing second year. Moderate use for pulp and firewood. Common. Dry sandy soils of scrub oak and pine barrens, well-drained sandy flatwoods, sandy river terraces. To about 200 m elevation. Mar–Apr. Syn: *Q. cinerea* Michx.

Identified by leaves widest at or near middle, the lower surface with stalked stellate hairs, the apex without a bristle tip; by gray bark. Trees to about 25 m tall by 80 cm DBH. Bark similar to that of *Q. alba,* scaly, sometimes becoming furrowed. First-year twigs with scattered stalked stellate hairs, often becoming glabrous. Leaves elliptic to narrowly elliptic, uncommonly obovate to narrowly obovate, margin entire, rarely some leaves with 1 to several humps or shoulders. Acorn cups turbinate, thin, sessile or on stalks to 1 cm long, covering ⅓ or a little more of acorn; scales small, appressed, tan to light chestnut, finely hairy. Acorns about 1 cm long. Infected by chestnut blight in some areas. Rare. Poorly drained soils that may become very dry during summer and fall; also on adjacent slopes. Piedmont of eGa and wSC; Caldwell Parish, La. Apr.

Identified by leaves widest at or near middle, the lower surface with a few to many stalked stellate hairs, the apex with a bristle tip which often drops with age (note photograph taken in late Sept has only one leaf with a bristle tip); by bark brownish. Trees to 25 m tall by 1.7 m DBH. Twigs glabrous, shiny. Leaf blades 2.5–4 × as long as wide, usually narrowly elliptic to elliptic, occasionally wider toward tip or base, 7–16 cm long, the tip pointed or rarely to obtuse. Acorn

cups 15–20 mm broad, enclosing ⅓ to ½ of acorn; scales finely hairy. Wood of moderate use as lumber and firewood; once used to make shingles, hence the common name. Planted as an ornamental. Common. Rich soils along streams, rich hillsides, dry uplands. Intolerant of shade. To 630 m elevation. Apr–May.

MORACEAE: Mulberry Family

3.
MACLURA. Osage-orange

Osage-orange;
Hedge-apple
Maclura pomifera (Raf.)
Schneid.
[206]

Identified by twigs bearing thorns that are sometimes quite scattered, with sap milky when evident, end bud an axillary one, stipules and stipule scars present; by fruits fleshy, 7–15 cm across, of many sections. Trees to 16 m tall by 2.3 m DBH. Inner bark and roots orange. Twigs with thorns that are solitary and simple, each beside an axillary bud; buds mostly hidden by bark; lenticels prominently elongated vertically. Leaves entire, glabrous, turning yellow in autumn, often clustered on stubby lateral branches as seen in photograph. Flowers tiny, unisexual, male and female on different plants, female flowers crowded in heads. Each fruit developing from many flowers, each section formed from a single flower. Mature fruits yellowish green, firm and fleshy, externally resembling an orange; sometimes eaten by livestock though may be poisonous. Wood bright orange, heavy, very tough and durable; used by Indians for bows, useful for fenceposts, excellent for firewood. Common. Usually in wet soils, but grows elsewhere. Probably native of csUS, but has been widely planted and escaping elsewhere. Apr–June.

4.
CUDRANIA. Cudrania

The related *Cudrania tricuspidata* (Carr.) Bureau ex Lavallee has the same identifying characters as *Maclura pomifera* except leaves sometimes 3-lobed toward end and fruits to 3.5 cm across, red at maturity. Rare escape from rare plantings.

EUPHORBIACEAE: Spurge Family

5.
SAPIUM. Sapium

Chinese Tallowtree
Sapium sebiferum (L.)
Roxb.
[207]

Recognized by leaf blades as wide as long or nearly so, a pair of glands on upper side of juncture of petiole and blade. Trees to 17 m tall by 97 cm DBH, rapid-growing, often forming thickets. Twigs glabrous, slender, brittle; sap milky and poisonous. Leaf tips acuminate, most of those in photograph worn off; leaves turning yellow to red in autumn. Flowers tiny, unisexual, in slender spikes to 10 cm long, male flowers toward the end, a few female flowers near the base. Fruits capsules, 10–18 mm across, slightly 3-lobed, the outer part splitting and falling off, leaving attached the 3 elliptical white waxy seeds. Planted as an ornamental, but easily spreading, often becoming weedy. In China the fruits are boiled in water and the wax extracted for use in making candles and soap. Occasional. Sandy soils in thin woods or in open. Mostly near the seacoast. Introduced from China and widely naturalized. A genus of about 100 species of temperate to tropical regions; only one native to N. Amer, in swAriz and nwMex. May–June.

ANACARDIACEAE: Cashew Family

6.
COTINUS. Smoketree

American Smoketree
Cotinus obovatus Raf.
[208]

Recognized by buds with acuminate and sharp-pointed scales; by pith of twigs continuous; by foliage and twigs with neither spicy-aromatic nor bitter-almond odor when crushed or broken; by leaves with silky hairs beneath when young, many or most of which fall with age, the apex obtuse to rounded. Shrubs or trees to 14 m tall by 48 cm DBH. Crown generally open, branches spreading and often drooping. Twigs slender, with prominent corky whitish lenticels; leaf scars with minute folds; terminal buds about 6 mm long. Leaves turning orange to scarlet in autumn. Flowers about 3 mm across; in clusters to 15 cm long, few per cluster and scattered; unisexual, the sexes usually on separate plants; petals greenish-white. Fruits dry 1-seeded drupes 3–6 mm long, on a slender stalk. Sometimes planted as an ornamental, but the only other species of the genus, *C. coggygria* Scop. of Eurasia, is more commonly used. Rare. Usually in limestone areas in dry places in open or in woods. Apr–May. Syn: *C. americanus* Nutt.

ELAEAGNACEAE: Oleaster Family

7.
ELAEAGNUS. Elaeagnus

Oleaster; Russian-olive
Elaeagnus angustifolia L.
[209]

Recognized by underside of leaves and first-year twigs with abundant scurfy scales; by stipules absent. Shrubs or trees to 14 m tall by 65 cm DBH, plants in the SE likely under half that size. Twigs often thorny, pith pale brown to orange-brown. Leaves lanceolate, 3–10 cm long, 1–3 cm wide, entire or rarely shallowly toothed, green to slightly silvery on the upper surface. Flowers fragrant, bisexual, short-pedicelled, in small lateral clusters, the ovary superior, appearing inferior in fruits. Fruits drupelike, actually each an achene enclosed by the enlarged persistent fleshy hypanthium; eaten by wildlife and seeds dispersed in their droppings. Used as an ornamental and planted in some land reclamation projects. Rare. Locally naturalized from plantings in the SE and westward to Cal and into Can from Ont to BC. A genus of over 40 species, all native to Eurasia except for a shrubby one that extends from the northern Great Plains into Alas. Apr–May.

LYTHRACEAE: Loosestrife Family

8.
LAGERSTROEMIA. Crepe-myrtle

Crepe-myrtle
Lagerstroemia indica L.
[210]

Recognized by first-year twigs winged or angled, at least below each leaf; by end buds axillary ones. Shrubs or trees to 10 m tall by 68 cm DBH. Leaves characteristically alternate, but opposite leaves may also be present to varying degrees. Flowers of early introductions usually pink; a variety of colors now available. Fruits nearly globose capsules 7–10 mm long. Rare. Sprouting from roots and spreading by seeds; sometimes forming colonies. A native of Asia often planted as an ornamental, escaping into a variety of habitats. June–Sept.

NYSSACEAE: Sourgum Family

9.
NYSSA. Sourgum; Tupelo

Recognized by twigs with 3 bundle scars per leaf scar, pith continuous and with firm diaphragms; by leaf blades longer than wide, the foliage lacking a spicy-aromatic or bitter-almond odor when crushed. Terminal buds present, with several scales. Leaves mostly entire, but sometimes with 1 to a few large irregularly spaced teeth, especially on sprouts and vigorous twigs. Flowers bisexual or unisexual, the sexes on the same or different trees. Fruits drupes; eaten by wildlife. Flowers a source of nectar for honey. A genus of 6 species, 2 in eAsia and 4 in eN. Amer.

KEY TO NYSSA SPECIES
1. Trees with 1–several trunks, these crooked; leaves mucronate and usually rounded at tip, mature ones velvety hairy below or rarely hairy only along veins; fruits 25–40 mm long, red when mature, on stalks 10–20 mm long, the stone paper-winged   1. *N. ogeche*
1. Trees with a single straight trunk; leaves rounded to pointed, not mucronate, mature ones glabrous or nearly so; fruits 10–40 mm long, blue to purple or

nearly black when mature, on stalks 15–55 mm long, the stone with indistinct
ribs to prominent sharp ridges   2
2. Leaf blades 13–30 cm long; mature buds globose to rounded; fruits 15–40
   mm long, usually one to a stalk, stone with sharp ridges   2. *N. aquatica*
2. Leaf blades to 15 cm long; mature buds ovoid; fruits 6–15 mm long, usually
   2 or more to a stalk, stone with rounded ridges   3
   3. Most leaves widest at the middle, the apex acuminate (rarely obtuse); fruits
      1–5 per stalk, stone faintly ribbed   3. *N. sylvatica*
   3. Most leaves widest beyond the middle, the apex obtuse (rarely some short-
      acuminate); fruits usually 1–2 per stalk, stone prominently ribbed
                                                                                       4. *N. biflora*

1.
Ogeeche Tupelo;
Ogeeche-lime
Nyssa ogeche
Bartr. ex Marsh.
[211]

Recognized by 1–several trunks, these crooked; by
leaves mucronate and usually rounded at tip; by fruit
25–40 mm long, red when mature, the stone with
10–12 papery wings. Shrubs or trees to 18 m tall by
77 cm DBH. Twigs hairy; buds nearly globose, obtuse,
terminal ones about 6 mm long, lateral ones much
smaller. Leaves velvety hairy beneath, uncommonly
hairy only along veins. Fruits with thick juicy sour
pulp; often remaining attached after leaves fall; can be
made into jam or jelly. Wood of poor quality and little
used. Occasional. Wet habitats; borders of streams,
ponds, and lakes, often leaning over the water; swamp
forests. Sprouting from roots as well as reproducing by
seeds, sometimes in dense stands. Apr–May.

2.
Water Tupelo
Nyssa aquatica L.

Similar to *N. ogeche,* especially in having large leaves,
13–30 cm long; large fruits, 15–40 mm long, and
usually 1 per stalk. It may be separated by the single
straight trunk; by leaves usually acuminate at apex,
often with large irregularly spaced teeth; by fruits dark
purple, the stone with sharp ridges. Trees to 32 m tall
by 2.6 m DBH. Wood soft, used for lumber and pulp.
Occasional. Swamps, floodplains, margins of ponds
and lakes. Apr–May.

3.
**Black Tupelo;
Blackgum**
Nyssa sylvatica Marsh.
[212]

Recognized by leaf blades to 15 cm long, mostly wid-est at the middle, apex acuminate or rarely to obtuse; by fruits 1–5 to a stalk, stone slightly 10–12-ridged. Trees to 42 m tall by 1.6 m DBH with a single trunk. Twigs glabrous or hairy; end bud a terminal one, ovoid, acute, about 6 mm long; 3 bundle scars per leaf scar. Leaves green beneath, glabrous or hairy, occasion-ally with 1–3 prominent irregularly spaced teeth. Fruits dark blue, 7–13 mm long, with a thin bitter usually sour pulp. Wood soft and tough, little used for lumber, occasionally used for pulp. Where flowers and fruits are lacking, Black Tupelo is easily confused with *Diospyros virginiana*. In the latter the leaves are whitish beneath, end bud an axillary one, and there is one bundle scar per leaf scar. Common. In thin woods or open, rarely in dense woods; usually in well-drained soils. Also in seMex. Mar–June.

4.
**Swamp Tupelo; Swamp
Blackgum**
Nyssa biflora Walt.
[213]

Recognized by leaf blades to 14 cm long, mostly wid-est beyond the middle, the apex obtuse to rarely short-acuminate; by fruits 1–2 to a stalk, stone prominently ribbed. Shrubs or trees to 34 m tall by 1.2 m DBH; sometimes in dense colonies. Twigs usually glabrous; terminal buds ovoid, acute, about 6 mm long. Leaves usually glabrous, margin lacking teeth. Fruits dark blue, 8–13 mm long, with a thin bitter usually sour pulp; eaten by wildlife, especially by waterfowl after fruits drop. Wood of similar quality and uses as that of *N. sylvatica*. Easily confused with *Diospyros virginiana;* see *Nyssa sylvatica* for separating characters. *N. biflora* is sometimes treated as a variety of *N. sylvatica;* exten-sive studies are needed to establish the proper rela-tionship. Common. Swamps, bogs, bottomlands, pond and lake margins. Apr–June.

CORNACEAE: Dogwood Family

10.
CORNUS. Dogwood

Alternate-leaf Dogwood
Cornus alternifolia L. f.
[214]

Recognized by pith continuous and without di-aphragms; by internodes decidedly unequal; by buds with 2, rarely 3, nearly equal scales; by alternate leaves, the blades considerably longer than wide, main

lateral veins attached to basal ⅔ of midrib only and arching in a manner that the tips of the veins are nearer the midrib than their middles. Shrubs or trees to 9 m tall by 55 cm DBH with reddish-brown bark. Branches nearly horizontal and in tiers, twigs with rank odor when bruised, terminal buds present but new twigs arising from lateral buds. Leaves light-colored beneath, mostly crowded at ends of twigs and not obviously alternate except on long twigs. Leaf margin sometimes minutely toothed, petioles slender. Fruits drupes, when mature dark blue, nearly globose, 6–8 mm across, on red stalks, flesh bitter, containing 1–2 grooved stones; eaten by wildlife. Attractive as an ornamental, but uncommonly planted. Occasional. Understory plant in broadleaf or coniferous woods; moist to dry habitats. To over 1900 m elevation in sAppalachians. Extends to Nfld. May–July.

ERICACEAE: Heath Family

11.
ELLIOTTIA. Elliottia

Elliottia;
Southern-plume
Elliottia racemosa Muhl. ex
Ell.
[215]

Recognized by twigs and fresh foliage without bitter-almond odor when crushed; by terminal buds and the several scales acuminate, one bundle scar, pith continuous; by leaf apex acute. Shrubs or trees to 15 m tall by 22 cm DBH. Bark light-gray, thin. Leaves sparsely hairy beneath. Petals separate, 4 or uncommonly 5. Fruits nearly globose capsules 10–12 mm across. Rare. Moist sandy soils, sandy ridges, less commonly in rocky areas. To about 120 m elevation. Almost entirely in CP; confined to Ga. Only species of the genus. June–Aug.

EBENACEAE: Ebony Family

I2.
DIOSPYROS. Persimmon

Common Persimmon
Diospyros virginiana L.
[216]

Recognized by end bud an axillary one, one bundle scar, stipules and stipule scars absent, pith continuous but with firm diaphragms, sometimes partly chambered in older twigs; by leaves without scurfy scales. Trees to 40 m tall by over 1 m DBH. Bark patterned in small blocks by longitudinal and cross fissures. Buds with 2 scales, lateral buds closely appressed. Leaves whitish beneath, often with black blemishes on upper surface, as in photograph. Flowers on previous year's twigs, functionally unisexual, the sexes on separate plants, those with fertile pistils having sterile stamens or none. Corolla of fertile flowers 12–20 mm long, petals united, the lobes spreading. Fruits berries to 4 cm across, with 1–several large flat seeds, the pulp orange-colored, commonly sweet when ripe and very bitter when not ripe; on some trees always a little to very bitter. Wood used for golf-club heads, veneer, occasionally low grade lumber. Popular food for wildlife, especially opossum and deer. Pulp of the fruit used locally for puddings, cakes, and beer. When not in flower or fruit sometimes confused with *Nyssa sylvatica*, which is easily separated by the end bud being a terminal one and by having 3 bundle scars. Common. Dry to wet habitats, usually in open. Widely spread by seeds dropped by wildlife. A genus of about 175 species of tropical and subtropical regions of Asia, Africa, and the Americas. Two species of tree size occur in the US, the Common Persimmon and another in Tex and adj. Mex. Apr–June.

Group K

Leaves simple, alternate, deciduous, serrate to dentate or lobed, more than 2-ranked.

KEY TO GENERA
1. Lateral buds evident and stalked, scales somewhat valvate   2. *Alnus*
1. Lateral buds sessile, sometimes quite inconspicuous, scales various   2
  2. Pith chambered; lateral buds superposed, close together   13. *Halesia*
  2. Pith continuous; lateral buds solitary in axils except occasionally in *Ilex*   3
    3. Twigs long remaining green, about 8-ribbed; leaves with a few to several conspicuous teeth; without stipules (See Group H)   *Baccharis*
    3. Twigs often green when young, but not remaining green after first year, not ribbed; leaves finely serrate to dentate or lobed; with or without stipules   4
      4. End bud an axillary one; stem tip present as a stub and/or scar beside the live end bud   5
        5. Bud scale 1, caplike; lateral twigs usually easily breaking off   1. *Salix*
        5. Bud scales 3–6; lateral twigs not easily breaking off   6
          6. Stipules absent, bundle scar 1, stems unarmed; petals united; fruits capsules   12. *Oxydendrum*
          6. Stipules present, bundle scars 3, stems sometimes thorny; petals separate; fruits drupes   7. *Prunus*
      4. End bud a terminal one, sometimes small and indistinct or the scales hidden by hairs; no stem stub or scar beside end bud, any scar nearby being from an early falling leaf or stipule   7
        7. Bundle scar 1, conspicuously protruding; pith about half the width of year-old twigs, with a network of firm strands   11. *Clethra*
        7. Bundle scars 1 to many, but not protruding; pith very small to about half width of year-old twigs, lacking any network of firm strands   8
          8. Stipules absent, bundle scar 1   9
            9. Leaves unlobed, bud scales indistinct and hairy. Apparently long extinct in the wild   10. *Franklinia*
            9. Leaves with 1 or 2 large entire lobes, bud scales evident and glabrous (See Group J)   *Sassafras*
          8. Stipules present, bundle scars 1–many   10
            10. Fresh twigs usually with a bitter-almond taste and odor when crushed or broken; lenticels long horizontally on vigorous twigs and branches over 5 mm across   7. *Prunus*
            10. Fresh twigs without this taste and odor; lenticels circular or indistinct on such twigs and branches   11

11. Leaf blades about as wide as long, apex acuminate (See Group G) *Populus*

11. Leaf blades longer than wide and usually not acuminate at apex 12

    12. Fruits drupes 10–30 mm across, with 1 stone
          7. Prunus

    12. Fruits not drupes or, if so, under 10 mm across or with 1–10 seedlike stones 13

13. Bundle scar 1; stipules very small, soon turning dark, persistent, sharp-pointed, being most common at juncture of current and last year's twig *8. Ilex*

13. Bundle scars 3 or more; stipules usually falling early and without the other above combination of characters 14

    14. Thorns always absent; 2–4 leaves and 2–5 buds at tip of vigorous twigs
          3. Quercus

    14. Thorns present or absent; one leaf and 1–2 buds at tip of vigorous twigs 15

        15. Thorns absent; lateral veins on each side of leaf evenly spaced and evenly sized except for gradual reductions toward the base and tip; ovary superior; fruits drupes with 1–4 bony stones *9. Rhamnus*

        15. Thorns absent or present; lateral veins not spaced and sized as above; ovary inferior; fruits pomes, sometimes with 1–5 bony stones 16

            16. Thorns usually present; fruits with 1–5 bony and usually 1-seeded stones *6. Crataegus*

            16. Thorns usually absent; fruits with 5 cavities with parchmentlike to firm walls, each cavity with 0–2 seeds 17

                17. Fruits with gritty masses in flesh, tapered to rounded at base, depressed at apex; leaves rolled inward in the bud, glabrous beneath when mature *4. Pyrus*

                17. Fruits with uniform flesh, depressed at both ends; leaves rolled or folded lengthwise in the bud, glabrous or hairy beneath when mature *5. Malus*

SALICACEAE: Willow Family

1.
SALIX. Willow

A primary recognition characteristic of most species of willows is the tendency for lateral twigs a year old or less to snap off easily when hit near the base. Confusion on this feature with any similar species is unlikely. Willows may be more positively identified by continuous pith in twigs, buds sessile, bud scale 1, and end bud an axillary one. Willows are fast-growing, most species having a relatively short life span, and the wood is soft and light. Bark aromatic and bitter-astringent, the bitterness probably due to tannic acid. Stipules usually present, dropping early in some species, sometimes absent; stipule scars small and occasionally absent. Flowers tiny, each sitting at the base of a scale, lacking sepals and petals, unisexual on separate plants, in both sexes arranged in conspicuous compact usually erect many-flowered cylindrical catkins that appear with or before the leaves, seldom later. Fruits 1-celled capsules that split at maturity into 2 parts; seeds minute, surrounded by long silky hairs. Various parts of the plant eaten by wildlife, especially the early spring leaves; the pollen eaten by bees. The slender flexible branches of some species are used locally in basket weaving. In the past, powdered bark, leaves, and buds, or extracts from them, were used medicinally for the salicin content, a name derived from *Salix*. Salicin has some of the same effects as aspirin (salicylic acid its main component), possibly altering to salicylic acid when eaten. Several species are planted as ornamentals, but most of these do not reproduce either sexually, because they are unisexual and rarely can produce seeds, or vegetatively, as necessary conditions are usually absent. Many native willows, in addition to reproducing by seeds, propagate profusely from easily detached twigs that root readily under moist conditions, populating stream banks, terraces, and sand or mud bars in and along streams. Species recognition sometimes depends on minute diagnostic characters provided by flowers and/or fruits which are available only for a short period in the spring. These lacking, precise identification may be difficult or impossible. Hybridization compounds the problem, as does the variability of some species from one part of their range to another, from one plant to another nearby one, and even on different parts of the same plant. The total number of *Salix* species has been estimated variously from around 300 to 500, most of them in the N. Hemisphere, largely in temperate and arctic zones. About 90 species are native to continental US, about 25 reach tree size, and 6 of these occur in the SE. In addition, there are 5 introduced tree-size species which have escaped and become naturalized. Of these 11 naturally reproducing species in the SE, only 3 are common, *S. caroliniana, S. nigra,* and *S. sericea,* the last uncommonly of tree size. These three and *S. babylonica,* an introduction commonly used as an ornamental, are illustrated and described; only brief information is provided for the other seven. Two keys are presented, one to the 3 common species and one to all species. The first key is easy to use and accurate for the 3 common species involved, and reasonable confidence can be placed in determinations made with it since the 8 rare species are unlikely to be encountered. The key to all species provides as many easily available and interpreted characters as practical but is more difficult to use and less accurate. Checking distributions can be a help. For this treatment we have depended on an extensive study of the genus by George W. Argus published in January 1986.

KEY TO COMMON SALIX SPECIES
1. Leaves about the same color below as above   1. *S. nigra*
1. Leaves lighter (whitish) below than above   2
    2. Mature leaves glabrous or with a few hairs beneath   2. *S. caroliniana*
    2. Mature leaves silky hairy beneath   3. *S. sericea*

KEY TO ALL SALIX SPECIES
1. Bud tip sharp-pointed, edges of bud scale free and overlapping, stamens 3 or
      more   2
    2. Underside of leaves green, not glaucous or rarely thinly so   1. *S. nigra*
    2. Underside of leaves glaucous   3

    3. Leaves elliptic to oblong, margin irregularly and shallowly toothed;
       capsules 6–7 mm long   5. *S. floridana*

    3. Leaves narrowly to broadly lanceolate, margin finely serrate; capsules 3–6
       mm long   4

        4. Stipules on vigorous twigs conspicuous; older twigs light brown, usually
           hairy; capsules 4–6 mm long   2. *S. caroliniana*

        4. Stipules small or absent; older twigs yellowish, tan, or grayish, usually
           glabrous; capsules about 3 mm long   6. *S. amygdaloides*

1. Bud tip blunt, edges of bud scales fused; stamens 2 or more   5

    5. Leaves green or pale beneath, not glaucous; stamens 3 or more   6

        6. Leaf tip long-acuminate; stipules persistent and prominently glandular on
           margin; capsules 5–7 mm long   7. *S. lucida*

        6. Leaf tip short-acuminate; stipules minute or absent; capsules 8–9 mm long
                                    8. *S. pentandra*

    5. Leaves glaucous beneath; stamens 2   7

       7. Leaf margin entire to finely crenate; capsules about 6 mm long
                                   9. *S. caprea*

       7. Leaf margin finely serrate to serrate; capsules 2.5–5 mm long   8

        8. Twigs and small branches greatly elongated and pendulous
                                  4. *S. babylonica*

        8. Twigs and small branches not greatly elongated and pendulous   9

           9. Leaves glabrous beneath   10. *S. fragilis*

           9. Leaves with dense silky hairs beneath   10

           10. Margin of largest leaves minutely serrate; male catkins on short
              leafy twigs; capsule tip acute; introduced   11. *S. alba*

           10. Margin of largest leaves serrate to minutely serrate; male catkins
              sessile; capsule tip blunt; native   3. *S. sericea*

1.
Black Willow
Salix nigra Marsh.
[217]

Recognized by tip of buds sharp-pointed, edges of bud scales free and overlapping; by leaves the same green color below as above. This is our most abundant willow and by far the most likely to be found having the underside of leaves green. Shrubs or trees to 42 m tall by 2.7 m DBH. Trunk usually leaning and often more than one; bark dark brown to blackish, ridged, furrowed, and scaly. Twigs brownish, slender, easily detached. The larger leaves 5–13 cm long, often slightly curved to one side; petioles 2–10 mm long. Flowers appearing with leaves. Fruiting catkins 2.5–7 cm long; capsules reddish-brown, glabrous, 3–5 mm long, pointed at tip. Important in reducing erosion of stream banks, bars, and islands. The wood with numerous uses including furniture, doors, cabinetwork, toys, crafts, and pulp. Once used as a source of charcoal for gunpowder. Birds eat buds and flowering catkins, deer the twigs and leaves, rodents the bark and buds. *S. gooddingii* Ball of swUS is sometimes combined with *S. nigra*. Common. Stream banks and terraces, bottomlands, pond and lake margins, depressions, and other moist to wet places. Reaches 1830 m elevation in Appalachian Mts. Extends into NB, sOnt, and e part of wTex. Feb–May.

2.
Coastal Plain Willow;
Carolina Willow
Salix caroliniana Michx.
[218]

Recognized by tip of buds sharp-pointed, edges of bud scales free and overlapping; by stipules large and roundish on vigorous twigs; by leaves lanceolate, glaucous beneath, marginal teeth evenly spaced; by capsules 4–6 mm long, glabrous. Other than *S. nigra*, the most likely tree-sized willow to be found, especially at low altitudes. Shrubs or trees to 25 m tall by 30 cm DBH. Bark gray to brown or nearly black, at first fairly smooth, later ridged and furrowed. Twigs slender, fragile. Leaves finely serrate, usually hairy when young, sometimes permanently hairy. Flowers appearing with the leaves, the catkins 3–9 cm long, on leafy stalks 2–5 cm long. Capsules narrowly lance-ovoid, 4–6 mm long, finely roughened, pointed at tip. Plants usually too small for use as lumber. Wildlife eat buds, catkins, leaves, twigs, and bark. There is some intergradation with *S. nigra*. In addition to leaf color, *S. caroliniana* may be distinguished by tiny yellowish glands on tips of the teeth or below them, whereas *S. nigra* has reddish glands at same locality. *S. caroliniana* and *S. amygdaloides* are sometimes confused. Twigs of the latter are yellowish to grayish, usually tenacious, and stipules, if present, minute. Twigs of the former are dark to light brown, fairly easily detached, and stipules on vigorous twigs roundish and conspicuous. Common. Stream banks, ditches, shores of ponds and lakes, depressions between dunes, marshes, swamps. Sometimes called Swamp Willow. To about 600 m elevation. Also occurs in Guatemala and wCuba. Feb–June. Syn: *S. longipes* Shuttlew.

3.
Silky Willow; Satin Willow
Salix sericea Marsh.
[219]

Recognized by tip of buds blunt, edges of bud scale fused; leaves narrowly lanceolate to lanceolate, with dense silky hairs on underside, these lying close to surface; by male catkins sessile; by capsule tip blunt. The most likely willow to be seen having silky hairs on leaf underside. Shrubs or rarely trees to 7 m tall by 12 cm DBH, usually with more than one stem, often forming thickets. Twigs dark brown to purplish, limber, breaking off easily at base. Leaves 3–14 cm long, finely serrate, whitish beneath, blackening on drying. Catkins appearing before leaves; capsules 3.5 mm long, obtuse at tip. Buds and flowers eaten by birds; leaves, twigs, and bark by various mammals. Common as shrub, rare as tree. Stream banks, sometimes in running water, seepage areas, bogs, marshes, ditches. To over 1600 m in sAppalachians. Mar–May.

4.
Weeping Willow
Salix babylonica L.
[220]

Perhaps the best known of the cultivated willows. Recognized by twigs and small branches much elongated and pendulous; by leaves glaucous beneath; by capsules 2.8–3.8 mm long. Trees to 32 m tall and 2.2 m DBH in cultivation. Bark rough, deeply furrowed. Twigs tough, yellowish. Leaves developing before those of the common Black Willow. Larger leaves 7–14 cm long, 9–18 mm wide, finely serrate, glaucous and usually silky hairy beneath, often nearly glabrous late in the growing season; petioles sticky, 7–12 mm long, glandular near base of blade. Catkins 18–35 mm long, appearing with the leaves. Capsules glabrous. Often used as an ornamental, most plants female and isolated from male ones. In the few instances where both sexes are together, reproduction by seeds can be conspicuous. Rare. To be expected as an escape throughout the SE; in parks, cemeteries, large estates, particularly in wet habitats such as margins of streams, ponds, and lakes. A native of China. Feb–Apr.

5.
Florida Willow
Salix floridana Chapm.

Shrubs, seldom trees, to 6 m tall by 10 cm DBH. Bud tip sharp-pointed. Leaf margin irregularly and shallowly serrate; blades elliptic to oblong, tip acute, underside glaucous and hairy. Recorded only in 10 counties: Pulaski and Early Cos. in sGa and in nFla from Jackson Co. nearly to cFla; possibly extinct in half of the counties. Rare. Along small streams, in wet limestone areas. Feb–Apr.

6.
Peachleaf Willow
Salix amygdaloides
Anderss.

Trees, rarely shrubs, to 20 m tall by 2.3 m DBH. Twigs yellowish; bud tip sharp-pointed. Leaves glaucous and usually glabrous beneath, shape similar to that of Peach; petioles 7–21 mm long. Stipules small or absent. Capsules about 3 mm long, glabrous, on stalks 1.6–3.2 mm long. Rare. Along streams, lake margins, low woods. Extends northwestward to eBC and scattered southward to swTex. Apr.

7.
Shiny Willow
Salix lucida Muhl.

Shrubs or trees to 18 m tall by 72 cm DBH. Bud tip blunt. Leaves green on both sides, shiny; petioles with glands near base of blade; stipules persistent and prominently glandular on margin. Capsules 5–7 mm long. Rare. Moist to wet places; meadows, along streams, lake margins, boggy places. Occurs north to Nfld and Lab; west to cSask and ND. Apr–May.

8.
Bay-leaved Willow
Salix pentandra L.

Shrubs or trees to 7 m tall. Bud tip blunt. Leaves resinous-fragrant, stipules minute or absent. Capsules 8–9 mm long. Rare. Spreading from cultivation locally. Native of Eurasia; in SE known as an escape only in wMd and Montgomery Co., Va.

9.
Goat Willow
Salix caprea L.

Shrubs or trees to 7 m tall. Bud tip blunt. Leaves with revolute entire to finely crenate margin, glaucous beneath, the largest ones 7–13 cm long, 2.5–5.5 cm wide, the widest likely to be encountered among the willows. Rare. Planted and escaping locally. Native of Eurasia; in SE known only from Baltimore Co., Md, and Macon Co., NC.

10.
Crack or Brittle Willow
Salix fragilis L.

One of the largest willows in the SE, to 34 m tall by 2.4 m DBH. Bud tip blunt. Leaves glabrous and glaucous beneath, coarsely and irregularly serrate, the teeth gland-tipped; petioles glandular near base of blade. Rare. Moist places. Native of Turkey, Iran, and Iraq. Cultivated, escaped, and well established locally.

11.
White Willow
Salix alba L.

Trees to 25 m tall by 2.7 DBH in cultivation. Bud tip blunt. Leaves glaucous and silky hairy beneath, margin of largest ones serrate to minutely serrate; petioles with tiny glands at junction with blade. Male catkins on short leafy twigs. Stamens 2. Capsule tip acute. Several varieties are known to have been planted in the SE; hybridization with 3 other species has been reported. Rare. Along streams and in wet soils, mostly in or near cities. Native of Eurasia; planted and escaped locally.

BETULACEAE: Birch Family

2.
ALNUS. Alder

Recognized by lateral buds evident and stalked, with 2–3 somewhat valvate scales. Twigs usually slightly zigzag. Major lateral veins of leaves evenly spaced, extending to the margin. Flowers small, unisexual, the sexes on the same plant. Male flowers in conspicuous dense pendulous cylindrical catkins; female flowers in much smaller catkins. Fruits nutlets with small wings, covered by scales arranged in conelike structures. Vegetative parts eaten by deer and beaver, the seeds by birds and rodents. A genus of about 30 species, primarily of north temperate regions of the world, but also some in more s areas at higher elevations. Eight species reach tree size in the US, 3 occurring in the SE.

KEY TO ALNUS SPECIES

1. Principal leaves with 5–8 pairs of lateral veins; flowering in late summer or early autumn, fruiting the second year; male catkins absent in winter   3. *A. maritima*
1. Principal leaves with 8–12 pairs of lateral veins; flowering in spring, fruiting same year; male catkins present in winter   2
   2. Leaves green on both surfaces, broadest at or beyond middle, finely and simply serrate, the teeth of nearly the same size   1. *A serrulata*
   2. Leaves a paler green or glaucous below, broadest at or below middle, doubly serrate with the teeth of irregular sizes   2. *A rugosa*

1.
Common Alder;
Hazel Alder
Alnus serrulata (Ait.)
Willd.
[221]

Recognized by principal leaves broadest at or beyond the middle, finely and simply serrate, the teeth nearly equal in size, green or nearly so on both surfaces, with 8–12 pairs of lateral veins. Shrubs or uncommonly trees to 12 m tall by 14 cm DBH; sometimes branching near the ground. Bark dark gray to brown, smooth, with numerous lenticels. Flowering in spring. Male catkins conspicuous in winter, pendulous. Female catkins erect, 3–6 mm long, maturing in fall to reddish ovoid conelike structures 12–18 mm long that persist into the following growing season. Common. Moist to wet places; along streams, ponds, lakes; in swamps. To over 900 m elevation. Occurs in seNS. Treated as *A. rugosa* Koch in earlier American manuals.

2.
Speckled Alder
Alnus rugosa (Du Roi)
Spreng.

Also having leaves with 8–12 pairs of lateral veins, flowers in spring, and male catkins conspicuous in winter. It may be separated by the paler green to glaucous underside of the leaves and doubly serrate margin with irregularly sized teeth. Shrubs or trees to 17 m tall by 26 cm DBH. Rare. Wet soils along streams, in swamps. Extends to Nfld, NWT, ceBC. *A. incana* Moench in earlier American manuals.

3.
Seaside Alder
Alnus maritima Muhl. ex
Nutt.

Similar in general appearance to the above 2 species. It may be recognized by leaves having only 5–8 pairs of lateral veins; by flowering in late summer or early autumn; by lacking male catkins during winter. Shrubs or trees to 9 m tall by 10 cm DBH. Rare. Stream banks and swamps near e shore of Md near sea level and in scOkla at about 200 m elevation.

FAGACEAE: Beech Family

3.
QUERCUS. Oak

Oaks generally can be distinguished from other genera of Group K by having 2–4 leaves and 2–5 buds at the tip of vigorous twigs. Absolute recognition is assured by presence of the unique acorns and their cups. Most oaks have a deeply penetrating taproot. Pith usually 5-angled, bundle scars in the leaf scars about 12 and difficult to count, stipules small and early deciduous, leaving inconspicuous scars. Leaves all petioled. Flowers unisexual, appearing before or as leaves develop, both sexes on the same plant. Male flowers in slender naked drooping catkins that are solitary or in groups (see photograph of *Q. illicifolia*); pollen wind-blown and abundantly produced by all species. Female flowers sessile, solitary, or in few-flowered spikes, each flower subtended by many scales, these forming the acorn cup (see one-year-old immature acorns in photograph of *Q. illicifolia*). Stigmas 3-lobed, the ovules 6. Fruits nuts (acorns); seeds with a thin papery coat; several aborted ovules normally adhering to coat near base or tip of the seed. Oaks are a genus of about 500 species of trees and shrubs in the N. Hemisphere, exclusive of the Arctic; about half of these are in the New World, being most abundant in cMex and extending north into Can and south into Columbia. About 68 species reaching tree size are native to the US; 37 of these and one naturalized species occur in the SE. Of the 38 species in the SE, 5 are evergreen, or generally so, and are appropriately described and illustrated in Group H. Five other species have entire unlobed leaves, or largely so, and are treated in Group J. The remaining 28 species are included here in Group K. All 38 are included in the key that follows. Oaks are divided into 2 distinct types, *white oaks* and *red oaks*. There are several differences that can be useful in identification, and some of these are presented in the following chart.

WHITE

1. Bark rather soft, gray and scaly; rarely dark and deeply furrowed.
2. Leaf tips and any lobes lacking bristles.
3. Acorns maturing first season.
4. Inner surface of acorn shell glabrous.
5. Aborted ovules at base of seed.
6. Embryo of seed sweet or slightly bitter.
7. Cup scales usually thickened basally and usually loosely appressed at tips.
8. Roots developing from seed in the autumn.

RED

1. Bark rather hard, dark, and furrowed; rarely somewhat scaly.
2. Leaf tips and lobes usually with bristles (falling early in some species).
3. Acorns maturing second season, rarely the first.
4. Inner surface of acorn shell velvety hairy.
5. Aborted ovules near apex of seed.
6. Embryo of seed usually quite bitter.
7. Cup scales scarcely thickened basally and usually tightly appressed at tips.
8. Roots developing from seed in the spring.

Of the 38 species of oaks in the SE, 17 are *white oaks* and 21 *red oaks*. Oaks are extremely abundant in the SE and of immense value economically and environmentally. Acorns are so important as food for wildlife that success or failure of an annual crop can be a major factor in winter survival. Some of the mammals that commonly include acorns in their diet are deer, hogs, bears, rats, squirrels, and raccoons; among birds are bluejays, woodpeckers, ducks, and wild turkeys. Foliage and twigs may be poisonous when eaten in considerable quantity. Oak lumber is hard, tough, durable, the porous wood stains well, and its uses are too extensive to enumerate. Several species of oaks are used as ornamentals.

Identification of Oaks

In naming oaks, utilize twigs with leaves from the exposed part of the crown when possible. Avoid heavily shaded leaves, twigs that have elongated a second time during the growing season, young trees, saplings, and rootsprouts. Look for acorns and their cups; if these are absent from the tree, look on the ground beneath the tree. Determining whether a tree is a *white oak* or a *red oak* can be helpful. Identification is complicated by the occurrence of hybrids, which often have quite variable leaves. All species are inclined to hybridize except that *white oaks* do not hybridize with *red oaks*. Since many hybrids lack fruit, it is a good idea to turn to nearby trees with fruit. When these fertile trees have been named it is often possible to recognize a hybrid in respect to the two parents that are involved. Hybrids can often be recognized as hybrids if they cannot be "keyed out" successfully and/or do not match any photographs and accompanying descriptions.

The photographs and descriptions of the 28 species in Group K are divided into the two types, the *white* and the *red;* those numbered 1–13 are *white oaks* and those numbered 14–28 are *red oaks*. Within each type the species are arranged for easier recognition according to similarities and differences in leaf shape and not necessarily according to natural relationships. Characters for positive identification of all variants in a species are sometimes numerous and are not included in a description. However by checking the photographs and the characters presented, identification to species can be had with reasonable accuracy. In the following Key to All Species the more obvious characters are used when possible, only occasionally and necessarily are minute and/or hard-to-recognize characteristics used.

KEY TO ALL QUERCUS SPECIES

1. Majority of leaves clearly widest near the tip and, if lobed, not prominently so   2
   2. Leaves 7–25 cm long, 7–20 cm wide, the lower surface with rusty yellow
     hairs or scurfy   16. *Q. marilandica*
   2. Leaves 3–14 cm long, 1.5–10 cm wide, the lower surface glabrous or hairy,
     but hairs not as above   3
     3. Twigs of current year densely and finely hairy, the cuticle rarely visible
       beneath the hairs (See Group J)   *Q. chapmanii*
     3. Twigs glabrous or, if hairy, the cuticle readily visible   4
       4. Leaves grayish beneath   1. *Q. breviloba*
       4. Leaves green beneath   5
         5. Leaves leathery, margin revolute (See Group H)   *Q. myrtifolia*
         5. Leaves membranous, margin flat   6
           6. Leaves averaging over 4 cm wide, a bristle usually on leaf tip;
             twigs of current year hairy   15. *Q. arkansana*
           6. Leaves averaging under 4 cm wide, bristle on leaf tip usually
             falling early; twigs of current year glabrous   14. *Q. nigra*
1. Majority of leaves widest near the middle or, if widest beyond the middle, then
  conspicuously lobed or toothed   7
   7. Leaves with lobes or teeth, these evenly spaced, uniform in shape, nearly equal
     in size except gradual reductions toward tip and base   8
     8. Leaves with conspicuous teeth, each tipped with a prominent bristle
                                   18. *Q. acutissima*
     8. Leaves lobed or, if toothed, the teeth lacking bristles   9
       9. Peduncle of mature cups 2–7 cm long; lower surface of leaves with
         numerous tiny dense stellate hairs (rarely very few) and clusters of 2–8
         erect or nearly so hairs   10. *Q. bicolor*
       9. Peduncle of mature cups under 1.5 cm long; lower surface of leaves with
         either tiny dense stellate hairs or only clusters of erect to suberect hairs
                                           10

10. Underside of leaves appearing glabrous, but with tiny stellate hairs
with 6–16 rays　　　　　　　　　　　　　　　　13. *Q. prinoides*
10. Underside of leaves with hairs evident, solitary and/or in clusters of
2–8, not stellate　　　　　　　　　　　　　　　　　　11
　　11. Scales of cup free nearly to base; trunk light gray, ridged and
　　　　furrowed, scaly; hairs on underside of leaves both solitary and
　　　　clustered with a spread of about 0.15–0.5 mm
　　　　　　　　　　　　　　　　　　　　11. *Q. michauxii*
　　11. Scales of cup fused except tips, which are sometimes weakly
　　　　appressed; trunk brown to nearly black, bark tight, on older
　　　　trees deeply furrowed; hairs on underside of leaves in clusters of
　　　　2–5 with a spread of about 0.1–0.25 mm　　12. *Q. montana*
7. Leaves entire or, if toothed or lobed, these not evenly spaced and/or not
uniform in shape　　　　　　　　　　　　　　　　　　12
　12. Leaves entire, not lobed or toothed, widest at or near the middle　　13
　　13. Leaves evergreen, underside grayish or light tan, often appearing
　　　　smooth, but densely covered with tiny stellate appressed hairs having
　　　　7–18 rays, with perhaps a few short erect hairs present　　14
　　　　14. Lower surface of leaves with appressed stellate hairs only; lateral
　　　　　　veins not deeply impressed in upper surface; leaf margin flat to
　　　　　　curved but not hiding any of leaf surface (See Group H)
　　　　　　　　　　　　　　　　　　　　　　Q. virginiana
　　　　14. Lower surface with appressed stellate hairs and a few short raised
　　　　　　hairs; lateral veins strongly impressed in upper surface; leaf
　　　　　　margin rolled to the extent that a narrow portion of the lower
　　　　　　surface is hidden (See Group H)　　　　*Q. geminata*
　　13. Leaves deciduous or evergreen, underside green to light tan, glabrous
　　　　or hairy, but hairs not both stellate and appressed except in the
　　　　deciduous No. 1　　　　　　　　　　　　　　15
　　　　15. Underside of leaves with abundant clusters of hairs that are fused
　　　　　　by their lower portions and spreading in the upper　　16
　　　　　　16. Leaves without a bristle tip; bark gray and like that of
　　　　　　　　Q. alba; mature acorns on twigs of current year (See Group J)
　　　　　　　　　　　　　　　　　　　　　Q. oglethorpensis
　　　　　　16. Leaves at first with a bristle tip that may later drop off; bark
　　　　　　　　dark; mature acorns on previous year's twigs　　17
　　　　　　　　17. Leaves mostly 2.5–4 × as long as wide (See Group J)
　　　　　　　　　　　　　　　　　　　　　Q. imbricaria
　　　　　　　　17. Leaves mostly 5–8 × as long as wide (See Group J, an
　　　　　　　　　　uncommon form)　　　　　　　*Q. phellos*
　　　　15. Underside of leaves glabrous or hairy, the hairs of any cluster free
　　　　　　of each other　　　　　　　　　　　　　18
　　　　　　18. Undersurface of leaves appearing smooth, but with minute
　　　　　　　　clusters of closely appressed stellate hairs, these sometimes
　　　　　　　　few; no bristles on leaf tip; mature acorns on twigs of current
　　　　　　　　year　　　　　　　　　　　　1. *Q. breviloba*

18. Underside of leaves glabrous except perhaps hairs in axils of lateral veins or beside midvein; bristles sometimes present on leaf tip; mature acorns on previous year's twigs 19

   19. Apex of most leaves obtuse to rounded, upper surface generally dull (See Group H) *Q. laurifolia*

   19. Apex of most or all leaves acute, upper surface shiny or dull 20

   20. Leaves thick and coriaceous, shiny above, vein network prominent above, evergreen most winters, rarely bilaterally symmetrical (See Group H) *Q. hemisphaerica*

   20. Leaves thin and pliable, dull above, vein network above indistinct, deciduous, bilaterally symmetrical (See Group J) *Q. phellos*

12. Leaves toothed or lobed, widest at or about middle or toward apex 21

   21. Majority of leaves with 0–4 bristles on margin, mature acorns on current or previous year's twigs 22

   22. Leaves averaging under 5 cm wide 23

   23. First-year twigs finely hairy 24

   24. Leaves glabrous beneath (See Group J) *Q. chapmanii*

   24. Leaves hairy beneath 25

   25. Leaves 3 × or more as long as wide (See Group J) *Q. incana*

   25. Leaves under 3 × as long as wide 26

   26. Leaves coriaceous, with no bristles on margin or at tip (See Group H) *Q. virginiana*

   26. Leaves pliable, usually with bristles on tip and on lateral lobes 17. *Q. ilicifolia*

   23. First year twigs glabrous 27

   27. Leaves with many clusters of radiating grayish hairs on the undersurface (See Group J) *Q. incana*

   27. Leaves glabrous beneath or with minute sessile clusters of horizontally spreading hairs 28

   28. Majority of leaves with acute tips (See Group H) *Q. hemisphaerica*

   28. Majority of leaves with obtuse tips 29

   29. Bark dark-colored; mature acorns on previous year's twigs; tip of leaves with a bristle that is often soon deciduous (See Group H) *Q. laurifolia*

   29. Bark light-colored; mature acorns on twigs of current year; leaves always without bristle tip 30

   30. Underside of leaves appearing smooth, but with minute clusters of closely appressed hairs 1. *Q. breviloba*

30. Leaves glabrous beneath   31
    31. Acorn cups flat at base, covering less
    than ¼ of acorn, scales glabrous except
    ciliate on margin   2. *Q. sinuata*
    31. Acorn cups rounded at base, covering
    ¼ to ½ of acorn, scales hairy near tip
    3. *Q. austrina*
22. Leaves averaging over 5 cm wide   32
    32. Undersurface of leaves pale to pale green; bearing numerous
    minute sessile stellate hairs with horizontally spreading rays
    (difficult to see, especially late in the year)   33
       33. Peduncles of acorn cups longer than the petioles; upper
       scales of cup long-acuminate   10. *Q. bicolor*
       33. Peduncles of acorn cups shorter than to about as long as
       petioles, upper scales acute to obtuse or tapering into
       long tips   34
          34. Lobes of the apical third of leaf as long or nearly so
          as those of the lower half, tip of lobes abruptly
          pointed; upper scales of acorn cup acute, not awned
          7. *Q. lyrata*
          34. Lobes of the apical third of leaf shorter than those of
          the deeply lobed lower portion, tip of lobes rounded;
          upper scales of cup long-awned   9. *Q. macrocarpa*
    32. Undersurface of leaves green, light green, or yellow-green;
    glabrous or with scattered to abundant hairs, these not sessile
    and stellate with horizontally spreading rays; most or all hairs
    clustered, in some plants fused at the base and tips spread-
    ing   35
       35. First-year twigs densely covered with fine grayish or
       tawny hairs that hide the cuticle in at least the tip tenth
       of the twigs   5. *Q. stellata*
       35. First-year twigs glabrous or with scattered to abundant
       hairs but the cuticle always visible beneath   36
          36. Leaves glabrous beneath   37
             37. Cup covering about ⅔ or more of acorn, tip of
             leaf lobes usually abruptly pointed   7. *Q. lyrata*
             37. Cup covering less than ½ of acorn, tip of lobes
             of leaf rounded   38
                38. Terminal buds ovoid, pointed; leaves 7–15
                cm long, 3–8 cm wide, with 3–7 lobes
                3. *Q. austrina*
                38. Terminal buds rounded, globose to broadly
                ovoid; leaves 10–20 cm long, 5–10 cm
                wide, with 7–11 lobes   8. *Q. alba*

36. Leaves hairy beneath, hairs may be few by autumn
39

   39. Hairs clustered, the clusters with hairs fused at
base and tips spreading   4. *Q. margaretta*

   39. Hairs clustered, none or rarely a few hairs with
bases slightly fused   40

      40. Cup covering about ⅔ or more of acorn
7. *Q. lyrata*

      40. Cup covering about half or less of acorn
6. *Q. similis*

21. Majority of leaves with 5 or more bristles on margin, mature acorns
on previous year's twigs   41

  41. Mature leaves with whitish, grayish, or yellowish hairs beneath,
these sometimes sparse late in year   42

    42. Leaves with small tawny hairs beneath; acorn cups turbinate,
the tips of marginal scales loose, forming a fringe
19. *Q. velutina*

    42. Leaves with whitish or grayish hairs beneath; acorn cups
shallow and not turbinate, the marginal scales closely
appressed   43

      43. Most leaf bases rounded   20. *Q. falcata*

      43. Most leaf bases broadly wedge-shaped   44

        44. Leaves under 12 cm long, mostly 5-lobed, underside
with dense grayish hairs   17. *Q. ilicifolia*

        44. Leaves over 10 cm long, lobes 5−11, the underside
with yellowish-gray hairs   21. *Q. pagoda*

  41. Mature leaves glabrous beneath, except sometimes with tufts of
hairs in axils of lateral veins   45

    45. Longest petioles under 18 mm long; scales along rim of cup
arched down about halfway inside the cup   22. *Q. laevis*

    45. Longest petioles 10−65 mm long; all cup scales remaining
outside of cup except often turning slightly in No. 23   46

      46. Midrib of upper leaf surface finely hairy, these often
scattered; largest terminal buds 6−12 mm long, the
scales hairy on the back (at least the upper ones); tip of
cup scales around rim loose, forming a fringe
19. *Q. velutina*

      46. Midrib of upper leaf surface glabrous; terminal buds 9
mm long or less, the scales hairy or glabrous on back,
sometimes with ciliate margin; tip of cup scales tightly
appressed   47

        47. Longest lateral lobes of leaves shorter than or about
equaling width of the undivided mid-section of leaf
23. *Q. rubra*

47. Longest lateral lobes of leaves obviously longer than
the width of the undivided mid-section of the leaf
48

48. Larger lateral lobes of most leaves with one
bristle per lobe, rarely 2 on some lobes
24. *Q. georgiana*

48. Larger lateral lobes of most leaves with 2 or
more bristles   49

49. Base of acorn cups strongly rounded to hemispherical or turbinate   50

50. Acorns with one or more small concentric grooves near the apex; scales of
mature cups light to dark reddish brown, glabrous or nearly so
25. *Q. coccinea*

50. Acorns without concentric grooves around apex; scales of mature cups ashy-
colored, hairy   51

51. Cups deeply turbinate, covering ⅓ to ½ of acorn   26. *Q. nuttallii*

51. Cups with rounded base, covering ¼ to ⅓ of acorn   28. *Q. shumardii*

49. Base of acorn cups flattened or nearly so   52

52. Cups 10–16 mm wide, acorns 10–15 mm long   27. *Q. palustris*

52. Cups 18–31 mm wide, acorns 15–37 mm long   28. *Q. shumardii*

1.
Shallow-lobed Oak
Quercus breviloba (Torr.)
Sarg.
[222]

Recognized by twigs glabrous or with scattered hairs; by leaves mostly 3–8 cm long and 2–4.5 cm wide, tip broadly rounded and bristleless, underside often appearing glabrous, but bearing numerous minute sessile stellate hairs with horizontally spreading rays; by acorn cups covering ¼ to ⅓ of acorn. Trees to 12 m tall by 81 cm DBH. Trunk with flaking gray bark. Leaves narrowly obovate to oblanceolate or narrowly elliptic, usually broadest beyond the middle, base wedge-shaped to obtuse, margin with 1 to 5 short rounded lobes or undulations. Well-shaded leaves and others late in season may be essentially glabrous beneath. Mature acorns on twigs of current year; cups rounded at base, 8–11 mm broad, scales finely grayish hairy. Rare. Upland scrub and woods, often in limestone areas, ridges. Occurs in neMex. Apr–May. Syn: *Q. durandii* var. *breviloba* (Torr.) Palmer; *Q. sinuata* var. *breviloba* (Torr.) C. H. Mull.

2.
Bastard Oak
Quercus sinuata Walt.

Similar to *Q. breviloba,* but may be recognized by leaves to 12 cm long and 6 cm wide and glabrous beneath; by acorn cups flat at base, 10–14 mm broad, covering less than ¼ of acorn. Rare. Wooded slopes, stream bluffs, alluvial woods. Apr. Syn: *Q. durandii* Buckl. in part.

3.
Bastard White Oak
Quercus austrina Small
[223]

Recognized by first-year twigs glabrous; by leaves 7–15 cm long and 3–8 cm wide, widest toward tip, the underside glabrous, with 3–6 obtuse to rounded bristleless lobes; by cup covering ¼ to ½ of acorn. Trees to about 20 m tall by 80 cm DBH. Trunk gray, furrowed, scaly. Acorn cups rounded at base, 12–16 mm broad; scales thin, narrow, with fine grayish to tan-colored hairs. Larger leaf forms of *Q. austrina* and the smaller ones of *Q. alba* can be confused. Acorn cups will distinguish the two species, scales of the former being thin, and basal ones of the latter much thickened. Rare. Fine sandy loams and other rich soils of wooded uplands, slopes, bluffs, and well-drained part of bottomlands. Apr. Syn: *Q. durandii* var. *austrina* (Small) Palmer.

4.
**Shrubby Post Oak;
Sand Post Oak**
Quercus margaretta Ashe
[224]

Identified by twigs glabrous or with scattered clusters of hairs, the cuticle readily evident; by leaves 3–10 cm wide, underside with abundant clusters of 4–9 hairs with fused bases and spreading upper portions. Shrubs or trees to 22 m tall by 1 m DBH, sometimes forming clumps from root suckers. Twigs brown or uncommonly varying to gray. Acorn cups hemispherical, 12–18 mm broad, covering ¼ to ⅓ of acorn, sessile or on peduncles to 8 mm long; scales small, the tips barely free and finely hairy. Acorns 9–21 mm long, seed slightly bitter. Easily confused with *Q. stellata,* which flowers about the same time when in the vicinity and has similar trunk and leaves. *Q. stellata* is clearly distinguishable by having leaves usually larger though of similar shape, undersurface with clusters of 4–9 hairs but rarely a few fused at the base, and by twigs densely hairy. Common in CP, scattered in Piedmont. Well-drained sandy, occasionally loamy, soils. To about 250 m elevation. Apr. Syn: *Q. stellata* var. *margaretta* (Ashe) Sarg.

5.
Post Oak
Quercus stellata Wang.
[225]

Recognized by first-year twigs densely covered with minute grayish to tawny hairs that hide the cuticle in at least the tip tenth of the twigs; by leaves 5–10 cm wide, lobed, the lobe tips lacking a bristle, underside grayish-green to brownish. During summer these hairs become matted and often form a "crust" which may become cracked and fall off in small sections, thus exposing a portion of the cuticle. Characteristically, tiny almost black granules are scattered in the matted hairs, a feature apparently lacking in other oaks. Trees to 27 m tall by 1.4 m DBH. Slow-growing. Trunk gray to light reddish-brown, normally fissured, the ridges rounded and scaly. Leaves usually with 5 primary lobes, the two nearest the terminal lobe quite broad; undersurface with abundant to scattered clusters of 4–9 hairs, rarely a few clusters with hairs fused at base. Acorn cups sessile, 12–27 mm broad, covering about ⅓ of acorn; scales small, finely hairy, closely appressed, with tips sometimes free. Common. Mostly in dry upland areas, especially in poor soils, including rocky slopes and ridges. Not shade tolerant. A common constituent of tree border at edge of prairies. To 900 m elevation, rarely to about 1500 m in sAppalachians. Mar–June.

6.
Bottomland Post Oak;
Delta Post Oak
Quercus similis Ashe

Similar to *Q. stellata,* but the hairs on first-year twigs do not hide the cuticle and the two major lobes adjacent to the terminal lobe of the leaf are often not as broad as in *Q. stellata*. Some leaf forms are similar to some of *Q. lyrata*. Tree to 30 m tall by 1.4 m DBH. Occasional. Moist to wet areas near streams and bayous. Apr. Syn: *Q. stellata* var. *paludosa* Sarg; *Q mississippiensis* Ashe.

7.
Overcup Oak; Swamp White Oak
Quercus lyrata Walt.
[226]

Recognized by first-year twigs glabrous or, if hairy, the cuticle readily visible; by leaves averaging over 5 cm wide, apical half about as deeply lobed as basal half, tip of lobes usually abruptly pointed and lacking a bristle; by underside of leaves glabrous or with scattered clusters of hairs, or with numerous minute stellate hairs with horizontal rays; by cup covering about ⅔ to nearly all of acorn, scales acute. Trees to 38 m tall by 2.1 m DBH. Trunk gray to brownish gray, with irregular ridges or small plates, these usually scaly. Leaves 7–20 cm long, with 5–9 short to long lobes which are often ascending, the leaf base broadly wedge-shaped. Acorn cups thin, 15–25 mm broad, nearly sessile or with peduncles to 4 cm long. Acorns 12–25 mm long, nearly globose to broadly ovoid. Plants with undersurface of leaves densely hairy sometimes called *Q. lyrata* f. *lyrata* and those with only scattered hairs *Q. lyrata* f. *viridis* Trel. Common. Low wet soils; bottomlands, along streams and bayous. Shade tolerant. Mar–May.

8.
White Oak
Quercus alba L.
[227]

Recognized by glabrous twigs; by leaves averaging over 10 cm long, glabrous beneath, with 7–11 uneven lobes, tip of lobes rounded and lacking a bristle; by cup covering less than half of acorn. Trees to 35 m tall by 2.8 m DBH. Trunk light gray, sometimes tinged with red or brown, fissured and ridged to flaky or scaly. Terminal buds globose to broadly ovoid. Leaves 10–20 cm long, 5–10 cm wide. Acorn cups sessile or on short peduncles; scales appressed, the basal ones much thickened. Acorns 15–35 mm long; seed slightly bitter. Major source of wood for whiskey barrels; a diminishing use for clapboard shingles and woven baskets. Planted as shade tree, the lower limbs remaining alive when shaded. Common. Well-drained soils, upland or lowland; in pure stands or with other species of trees. From near sea level to 1800 m elevation, but largely below 1500 m. Abundant in sAppalachians. Apr–June.

9.
Bur Oak;
Mossy-cup Oak
Quercus macrocarpa Michx.
[228]

Recognized by leaves 5–15 cm wide, the lobes of the apical third shallower than those of the basal half; tip of lobes rounded and lacking bristles; underside pale, with numerous minute sessile stellate hairs with horizontally spreading rays, these hairs sometimes mostly shed by late in growing season; by scales of acorn cup long-awned and forming a prominent fringed border on the cup. Trees to 40 m tall by 2.6 m DBH, uncommonly shrubby. Trunk deeply furrowed, the ridges broken into thick elongate brown scales. Terminal buds globose to ovoid. Acorn cups 2–5 cm broad, thick, often covering half to most of acorn, rarely as little as a third. Common. Bottomlands to dry slopes and uplands, especially in limestone areas. To about 900 m elevation. Extends west into neWyo and seMont; north into seSask. Apr–May.

10.
Swamp White Oak
Quercus bicolor Willd.
[229]

Recognized by leaves 5–11 cm wide, shallowly lobed, lobes without bristles, underside pale with abundant (rarely few) minute sessile stellate hairs with horizontal rays and also with clusters of 2–6 larger nearly erect hairs; by acorns on peduncles 2–7 cm long that exceed the petioles. Trees to 38 m tall by 2 m DBH, rapid-growing. Trunk gray, deeply furrowed with elongate scaly ridges, these sometimes flat-surfaced. Terminal buds ovoid to subglobose, usually hairy near the tip. Leaves usually oblong-obovate to obovate, often unevenly lobed, lobes rounded to pointed and tipped with a small gland. Acorn cups thick, hemispherical or nearly so, covering ⅓ to ½ of acorn, scales acute to very long acuminate and often forming a fringed edge. Acorns mostly 2–3 cm long; seed sweet. Common. Bottomlands, swamps, floodplains, stream margins. Shade tolerant. To about 600 m elevation. Apr–May.

11.
Swamp Chestnut Oak;
Basket Oak
Quercus michauxii Nutt.
[230]

Recognized by having 9–14 parallel main lateral veins on each side of leaves, these evenly spaced and each terminating in a large rounded tooth, the teeth lacking bristles and all about the same size except gradually smaller toward the tip and base of leaf; by underside of leaves with hairs both solitary and in clusters of 2–8 with a spread of about 0.15–0.5 mm. Trees to 40 m tall by 2.2 m DBH. Trunk light gray, ridged and furrowed, scaly. Terminal buds mostly acute. Leaves

widest beyond the middle, to 28 cm long and to 16 cm wide. Acorn cups largely 25–35 mm broad, enclosing about ⅓ of acorn, sessile or on peduncles to 1 cm long; scales hairy, free nearly to their bases. Acorns 25–35 mm long; seeds sweet. Much like *Q. montana,* which differs by having a trunk brown to nearly black, bark tight, and hairs of underside of leaf only in clusters of 2–5 with a spread of about 0.1–0.25 mm. Some forms of *Q. prinoides* are similar to *Q. michauxii,* but the former can be recognized by the minute stellate hairs on the leaf undersurface and by acorn cup scales free only at tip. Common. Low areas, riverbottoms, swamp borders, in ravines in parts of distribution. Shade tolerant. To about 300 m elevation. Syn: *Q. prinus* L. Apr–May.

12.
Chestnut Oak
Quercus montana Willd.
[231]

Recognized by having 9–16 parallel main lateral veins on each side of leaves, these evenly spaced and each terminating in a large rounded tooth, the teeth lacking bristles and all about the same size except gradually smaller toward the tip and base of leaf; by underside of leaves with hairs only in clusters of 2–5 with a spread of about 0.1–0.25 mm. Trees to 30 m tall by 2.8 m DBH. Trunk brown to nearly black, bark tight, on the older trees deeply and coarsely furrowed, the ridges continuous. Terminal buds acute to acuminate. Leaves commonly widest beyond the middle, to 30 cm long and to 10 cm wide. Acorn cups sessile or on peduncles to 1 cm long; edges thin, covering about ⅓ of acorn; scales fused except at tip. Acorns largely 20–35 mm long; seed sweet. Much like *Q. michauxii,* which check for distinguishing characters. Somewhat like some forms of *Q. prinoides,* which can be recognized by the minute stellate hairs on leaf undersurface. Common. Ridges, dry mountain slopes, rocky hillsides, well-drained portions of lowlands. Not shade tolerant. To about 1500 m elevation. Apr–June. Syn: *Q. prinus* L. of some authors.

13.
Chinquapin Oak
Quercus prinoides Willd.
[232]

Recognized by leaves averaging over 5 cm wide, margin with 3–14 coarse pointed bristleless teeth on each side, these evenly sized and spaced except gradual reductions in size toward tip and base; underside often apparently glabrous, but with minute stellate hairs

with 6–16 horizontal rays and a spread of about 0.1–
0.2 mm; by peduncle of mature cups to 1 cm long,
shorter than petioles. Shrubs or trees to 24 m tall by
2.8 m DBH. Trunk ashy gray, sometimes tinged with
brown, fissured and flaking. Leaves 5–20 cm long, 3–
10 cm wide. Acorn cups 10–20 mm broad, covering
about ⅓ to ½ of acorn; scales small, fused or free only
at tip. Seed sweet. Wood heavy and durable. Some
authors consider plants with 3–8 pairs of lateral veins
that end in lobes as Dwarf Chinquapin Oak (*Q. prinoides*)
and plants with 9–15 pairs of lateral veins ending
in lobes as Chinquapin Oak (*Q. muehlenberggi* Engelm.),
the former a shrub, rarely a tree. Some forms of *Q. prinoides* are similar to *Q. montana* and *Q. michauxii* and
these should be checked for separation characters. Occasional. Dry sandy or rocky soils, clay soils of wooded
slopes, most frequent in limestone areas. To about 900
m elevation. Extends in scattered localities to seNMex,
and south into neMex. Apr–May. Syn: *Q. prinoides* var.
acuminata (Michx.) Gleason.

14.
Water Oak
Quercus nigra L.
[233]

Recognized by glabrous twigs; by leaves membranous,
averaging under 4 cm wide, widest toward tip which is
usually without a bristle as it falls off early. Trees to 32
m tall by 2 m DBH. Buds 4–7 mm long; scales dark
brown, finely hairy. Semi-evergreen in warmer parts of
range with leaves gradually dying during winter.
Leaves 4–10 cm long and 2–5 cm wide, gradually
narrowed to the base, margin without teeth or lobes,
or sometimes with 3 rounded to rarely pointed lobes,
with one bristle on tip of each at first. Leaves of sprouts
and twigs formed by a second growth of the season
often with additional lateral sharp-pointed lobes.
Acorn cups thin, 10–15 mm broad, or as small as 6 mm
broad in rare forms, flat (rarely) to strongly rounded
at base, covering ¼ to ⅓ of acorn. Copious crops of
acorns are usually produced every year or two. A rapid-
growing tree, frequently planted for shade, but of low
quality because of the susceptibility of older trees to
rot. Wood of only moderate quality for lumber. Common. Best adapted to poorly drained soils in woods or
open, but spreading to a variety of other habitats including well-drained soils. To about 300 m elevation.
Mar–May.

15.
Arkansas Oak
Quercus arkansana Sarg.
[234]

Recognized by twigs hairy, rarely nearly glabrous; by leaves membranous, averaging over 4 cm wide, green beneath, widest toward the tip, which usually bears a bristle. Trees to 24 m tall by 1.1 m DBH. Trunk often crooked. Current year's twigs usually with many stellate hairs; buds normally reddish-brown, the scales ciliate on margin. Leaves 5–14 cm long, 3–10 cm wide. Acorn cups 12–14 mm broad, with rounded base, enclosing about ¼ of acorn. Hybridizes especially with *Q. nigra* which flowers about the same time. Rare, being most abundant in swArk. In sandy and loamy sand soils, occasionally on rocky slopes, around heads of small creeks. Shade tolerant, frequently found under other hardwoods. Mar–Apr. Syn: *Q. caput-rivuli* Ashe.

16.
**Blackjack Oak;
Blackjack**
Quercus marilandica
Muenchh.
[235]

Recognized by leaves clearly widest near tip, 7–20 cm wide, lower surface with rusty yellow hairs or scurfy. Trees to 14 m tall by 1.3 m DBH. Trunk nearly black, divided into hard irregular to rectangular plates. Twigs stout, thinly hairy; terminal buds 5–10 mm long, angled, scales with rusty hairs. Leaves leathery, shiny above, with broad usually bristle-tipped lobes; the base rounded to subcordate. Acorn cups 15–20 mm broad, turbinate to nearly hemispherical, covering about half of acorn; scales blunt, hairy on back, forming a loose fringe at margin of cup. Acorns finely hairy. Common. Dry sandy and clay soils, rocky upland areas. Abundant at margins of woodlands adjoining prairies. Intolerant of shade. To about 900 m elevation. Mar–May.

17.
Bear Oak; Scrub Oak
Quercus ilicifolia Wang.
[236]

Recognized by first-year twigs matted with dark gray hairs; by leaves usually under 12 cm long, the bases angular, mostly 5-lobed, the lobes usually bristle-tipped, undersurface with gray felted hairs. Shrubs or trees to 8 m tall by 15 cm DBH. Terminal buds chestnut-brown, 3–4 mm long, blunt, ovate. Leaves ovate to elliptic or obovate, 5–13 cm long, 3–8 cm wide, with 3–7 short triangular to ovate lobes. Upper surface of young leaves with dense grayish hairs that

K

soon fall off; mature leaves dark green above. Acorn cups 10–16 mm broad, rounded at base, usually with a scaly stalklike neck, covering about half of acorn; scales thin, light brown. Immature one-year-old cups and acorns are shown in the photograph. Seeds very bitter. Often forming dense thickets, providing food and shelter for wildlife. Occasional. Poor soils of dry slopes, mountain ridges, sandy barrens. To over 900 m elevation. May–June.

18.
Sawtooth Oak
Quercus acutissima Carruth.
[237]

Recognized by leaves with evenly spaced lateral veins, these ending in conspicuous bristle-tipped teeth. Trees to about 20 m tall with a rounded crown. Bark furrowed. Buds acute, circular in cross-section. Leaves to 18 cm long and 5 cm wide, tip acuminate. Acorn cups with spreading long-tapered scales, covering about ⅓ of mature acorn. Native of Asia. Planted as an ornamental and as food for wildlife. Rare. Reproducing near plantings in several localities in SE. Apr.

19.
Black Oak
Quercus velutina Lam.
[238]

Recognized by leaves very shallowly to deeply 5–9-lobed with 9 or more bristles along the margin, midrib of upper surface finely hairy, the hairs often scattered and late in growing season reduced to mere rough spots; by tips of upper scales of acorn cups loose, forming a fringe around rim of cup. Trees to 35 m tall by 2 m DBH. Bark on older trees very dark, deeply furrowed, the ridges broken; inner bark yellow to orange, providing a natural yellow dye, very bitter. Terminal buds 6–12 mm long, pointed, strongly angled, the brownish scales with gray hairs. Lower leaf surface glabrous except for tufts of hairs in axils of main veins, or rarely hairy on entire undersurface. Acorn cups finely hairy on inner surface, sessile or nearly so, turbinate or nearly so, covering about ⅓ to ½ of acorn. Acorns 12–25 mm long; seeds bitter. Common. Dry upland areas, especially rocky or sandy soils, sometimes on well-drained terraces along streams. To about 1500 m elevation. Apr–May.

20.
Southern Red Oak
Quercus falcata Michx.
[239]

Recognized by leaves conspicuously U-shaped at base, 10–30 cm long, 8–13 cm wide, prominently 3–9 lobed, lobes often falcate, terminal lobe frequently strap-shaped and in one leaf form oblong-obovate with 3 shallow lobes at apex; by undersurface of leaves with dense soft minute gray to light rusty-gray hairs. Trees to 39 m tall by 2.7 m DBH. Trunk dark brown to nearly black, narrowly fissured between narrow flat ridges. First-year twigs reddish, finely hairy, sometimes becoming nearly glabrous. Terminal buds ovoid, pointed, reddish-brown, to 8 mm long; back of scales with brownish hairs. Acorn cups covering ⅓ to ½ of acorn, the base nearly flat to rounded and often narrowed to a stalklike base. Seeds bitter. Often planted as a shade tree. The similar *Q. pagoda* is easily separated by its broadly wedge-shaped leaf bases instead of rounded, and by lobes rarely falcate and more nearly at right angles than in *Q. falcata*. Common. Dry uplands, in sandy, loamy, or clay soils. In open or mixed with other trees. To about 750 m elevation. Mar–May.

21.
Cherrybark Oak
Quercus pagoda Raf.
[240]

Recognized by leaves broadly wedge-shaped at base, or nearly so, mostly 10–18 cm long and 7–12 cm wide, prominently 5–11-lobed, lobes scarcely or not at all falcate; by leaf undersurface with dense minute soft yellowish-gray to light brown hairs. Trees to 40 m tall by 2.8 m DBH. Trunk dark reddish-brown to nearly black, at first smooth then narrowly fissured between flat plates, the bark suggestive of that of *Prunus serotina*. In large trees the bark on the trunk up among the limbs displays grayish horizontal bands. First-year twigs reddish, usually with small dense rusty hairs that sometimes nearly obscure the cuticle, later most of the hairs dropping off. Terminal buds ovoid, pointed, reddish-brown, backs of scales hairy. Acorn cups covering ⅓ to ½ of acorn, the base nearly flat to rounded, but often narrow and stalklike. Seeds bitter. Considered by some authors a variety of *Q. falcata*. Although similar to this species in some respects, it has a longer branch-free trunk and has other characters which are contrasted under *Q. falcata*. The two species also largely occupy different habitats. Common. Better-drained portions of floodplains and bottomlands and their margins. To about 250 m elevation. Apr–May. Syn: *Q. pagodifolia* (Ell.) Ashe; *Q. falcata* var. *pagodifolia* Ell.

22.
Turkey Oak
Quercus laevis Walt.
[241]

Recognized by mature leaves 9–35 cm long, with 3–7 irregular lobes, margin with 7 or more bristles, underside glabrous except for tufts of hairs in axils of lateral veins, petioles under 18 mm long; by scales along rim of cup arched down into the cup about halfway. Trees to 25 m tall by 82 cm DBH. Trunk with bark thick, blackish, with deep irregular furrows and rough blocky ridges. Terminal buds to 12 mm long, pointed, narrowly ovoid, rusty hairy toward tip. Of the oaks having similar leaf forms, *Q. laevis* is the only one with petioles twisted so that the plane surfaces of the blade are nearly always perpendicular to the ground. Acorn cups 2–3 cm broad, turbinate, on peduncles to 5 mm long, covering about ⅓ of acorn. Seeds bitter. Common. Dry sandy areas, well-drained ridges. An important component of oak scrub areas. To 150 m elevation. Apr. Syn: *Q. catesbaei* Michx.

23.
Northern Red Oak
Quercus rubra L.
[242]

Recognized by leaves 13–20 cm long, 8–15 cm wide, with 7–11 lobes, the lobes at midleaf shorter than to about equaling the width of the undivided midsection of the leaf, most lobes with more than one bristle, upper and lower sides glabrous or occasionally with tufts of hairs in axils of the larger veins on the underside. Trees to 35 m tall by 2.2 m DBH. Trunk of young trees smooth, greenish-brown, later broken into wide nearly smooth elongate plates by shallow dark furrows. First-year twigs soon glabrous; terminal buds ovoid and pointed, back of scales glabrous. Leaves dull above. Acorn cups flat to shallowly turbinate or hemispherical, 18–30 mm broad, thick, sessile or short peduncled; scales tight, the ones along rim often turned slightly into the cup. Seeds bitter. Northern Red Oak is sometimes separated into two varieties, *rubra* and *borealis* (Michx. f.) Farw. In the former the cup covers about ¼ of the acorn, the leaf lobes usually cut ¼ the way to the midvein, or less. In the latter the cups cover about ⅓ of the acorn, the leaf lobes usually cut over ¼ the way to the midvein. Var. *borealis* occurs farther north and in the sAppalachians extends to higher elevations than does var. *rubra*. Common. Wooded mountain slopes and ravines; southward and at lower elevations more often in ravines or on north- and east-facing slopes. To 1800 m elevations in sAppalachians. Extends northeastward into NS. Apr–May. Syn: *Q. maxima* (Marsh.) Ashe.; *Q. borealis* Michx. f.

24.
Georgia Oak
Quercus georgiana
M. A. Curtis
[243]

Recognized by leaves 3–12 cm long, a majority with 5–6 lobes, almost all with one bristle per lobe, sometimes 2 on the larger leaves; by undersurface glabrous except for tufts of hairs in axils of main veins. Shrubs or trees to 14 m tall by 31 cm DBH; often quite stunted on granitic outcrops. Trunk dark gray, thin, and smooth, becoming scaly. Terminal buds ovoid to narrowly ovoid, dark brown. Leaves 2.5–9 cm wide, 3–6 lobed. Acorn cups 10–17 mm broad, covering ¼ or rarely to ½ of acorn, the base nearly flat to rounded, sessile or on peduncles to 5 mm long. Acorns 8–13 mm long. Seeds bitter. Rare. On and around granitic outcrops; rocky knolls, ridges, and slopes. To about 500 m elevation. Apr.

25.
Scarlet Oak
Quercus coccinea Muenchh.
[244]

Recognized by mature leaves 7–16 cm long, 8–13 cm wide, deeply 5–9-lobed, the sinuses rounded, larger lobes with 3 or more bristles, both sides of leaf shiny and glabrous except for small tufts of hairs in axils of main veins below; by acorn cups turbinate or nearly so, scales glabrous or nearly so; by acorns with one or more concentric grooves around tip. Trees to 29 m tall by 1.7 m DBH. Trunk dark gray to dark brown, irreg-, ularly and shallowly fissured into scaly ridges or narrow plates; lower branches horizontal or nearly so. Terminal buds 3–7 mm long; scales finely hairy, especially near tip. Acorn cups 15–22 mm broad, covering ⅓ to ½ of acorn, scales light to dark reddish-brown, triangular, appressed, glabrous or nearly so. Acorns 12–20 mm long. Seeds bitter. *Q. coccinea* is similar to *Q. palustris* in several aspects; characters to separate the two are given under the latter species. *Q. coccinea* also shares several characteristics with *Q. shumardii,* but in the latter leaves average larger, acorn cups 18–31 mm broad and with a rounded base, acorns 15–37 mm long. Common. Dry upland areas, especially on ridges and mountain slopes. To about 1500 m elevation. Apr–May.

26.
Nuttall Oak
Quercus nuttallii Palmer
[245]

Recognized by mature leaves 10–23 cm long, 5–12 cm wide, glabrous beneath except tufts of hairs in axils of main veins, deeply 5–9-lobed, larger lobes with 3 or more bristles, the sinuses rounded; by acorn cups deeply turbinate, the base stalklike and scaly; by acorns 20–28 mm long with brownish longitudinal stripes and lacking concentric grooves around apex. Trees to 40 m tall by 1.6 m DBH. Trunk gray to brownish, smooth, with age becoming dark and with scaly ridges. Twigs glabrous, usually brownish gray; buds 3–6.5 mm long, ashy gray to brownish gray, scales ashy-colored and hairy with margin usually ciliate. Acorn cups 15–20 mm broad, covering ⅓ to ½ of acorn; seeds bitter. Shares several features with *Q. coccinea* but the latter may be distinguished by acorns which are 12–20 mm long with one or more concentric grooves around the tip. *Q. nuttallii* is also similar in several respects to *Q. palustris* and *Q. shumardii*. For their separation turn to couplet 51 of Key to All Quercus Species. Common. Floodplains, bottomlands, river terraces. To about 150 m elevation. Apr–May.

27.
Pin Oak
Quercus palustris Muenchh.
[246]

Recognized by mature leaves 7–15 cm long, 4–10 cm wide, deeply 5–7-lobed, larger lobes with 3 or more bristles, the sinuses rounded, glabrous on both sides except small tufts of hairs in axils of main veins on underside; by acorn cups 10–16 mm broad, nearly flat at base, covering about ¼ of acorn; by acorns lacking concentric grooves around tip. Trees to 28 m tall by 2.2 m DBH. Trunk grayish-brown, bark very tight, only larger trees with wide furrows. Lower branches usually angled downward, their stubs remaining on trunk after the branches die. Terminal buds 3–6.5 mm long, ashy-gray to brownish-gray, scales glabrous to finely hairy toward tip. Acorns 10–15 mm long. Used as an ornamental. *Q. palustris* is similar to *Q. coccinea,* but in the latter lowest branches are horizontal or nearly so, cups cover ⅓ to ½ of acorn, and concentric groves are present around tip of acorn. *Q. palustris* also shares several characteristics with *Q. nuttallii,* but in the latter lobes of the leaf are generally wider, sinuses usually smaller, acorn cups are strongly turbinate and cover ⅓ to ½ of acorn. Common. Compact soils subject to flooding, poorly drained upland areas. To about 300 m elevation. Mar–May.

28.
Shumard Oak
Quercus shumardii Buckl.
[247]

Recognized by mature leaves 8–20 cm long, 6–15 cm wide, deeply 7–11-lobed, larger lobes with 3 or more bristles, the sinuses rounded and generally deep, glabrous on both sides except small tufts of hairs in axils of main veins on underside; by acorns lacking concentric grooves around tip. Trees to 40 m tall by 2.1 m DBH. Bark thick, smooth, and grayish, becoming dark gray and furrowed into low ridges. Lower branches chiefly horizontal. Terminal buds 3–5 mm long; scales straw-colored, usually smooth. Acorns 15–37 mm long. *Q. shumardii* is similar in several respects to *Q. coccinea* and *Q. nuttallii*. For their separation turn to couplet 49 of Key to All Quercus Species. Occasional. Rich woods, bottomlands, slopes, bluffs. Apr–June.

ROSACEAE: Rose Family

4.
PYRUS. Pear

Pear
Pyrus communis L.
[248]

Recognized by year-old twigs reddish-brown, lenticels circular or nearly so, end bud a terminal bud, lateral buds sessile, pith continuous; by leaves longer than wide, stipules or their scars present; by fruits pomes with gritty masses in the flesh. Trees to 20 m tall by 1 m DBH. Leaves shiny green and smooth on upper side. Flowers with an unpleasant odor. Fruit a pear to 10 cm long, with green to brown skin, maturing in late summer. Widely planted for its fruit. The fine-grained hard reddish-brown wood has been used in furniture, for carvings, and a variety of other items. Occasional. Planted throughout the SE, escaping to fencerows, old fields, edges of woods, waste places. Mar–May.

5.
MALUS. Apple; Crabapple

Recognized by twigs with continuous pith, 3 or more bundle scars, sessile lateral buds, and end bud a terminal one; by leaves longer than wide, stipules or their scars present; by fruits pomes, depressed at both ends, without gritty masses in the flesh. Stems sometimes thorny; twigs usually with licorice flavor. Inflorescence a few-flowered cluster. Flowers with 5 sepals, 5 petals, numerous stamens, 5 carpels that are united at their bases but with 5 separate styles, and the 5 united ovaries surrounded by and fused to the receptacle. Seeds enclosed within parchmentlike to firm walls developed from the ovary walls; flesh of the fruits derived largely from the receptacle. Fruits eaten by wildlife. Leaves and fruit frequently infected by apple-cedar rust. Spores produced by this rust on *Malus* can infect only *Juniperus* species and the spores on *Juniperus* can infect only *Malus*. See photograph of *M. ioensis* for appearance of the rust. Some Red Haws, *Crataegus* spp., are similar to some *Malus* spp., but differ in that the ovary walls become bony in *Crataegus* fruits, and branches almost always are thorny. *Malus* is sometimes included under the genus *Pyrus*.

KEY TO MALUS SPECIES
1. Leaves permanently hairy beneath, sepals hairy on outside   2
    2. Leaves with uniform fine teeth, not lobed, rolled inward in the bud
            1. *M. pumila*
    2. Leaves coarsely serrate, some usually lobed, folded in the bud   2. *M. ioensis*
1. Mature leaves glabrous beneath (may be quite hairy when young), sepals glabrous on outside   3
    3. Leaves of fertile twigs, and usually the others, mostly crenate-serrate, rounded to obtuse to barely acute   3. *M. angustifolia*
    3. Leaves of fertile twigs, and usually the others, all serrate to doubly serrate, often some with triangular lobes at widest part, mostly acute to acuminate
            4. *M. coronaria*

1.
Apple
Malus pumila Mill.
[249]

Recognized by leaves permanently hairy on underside, with uniform fine teeth, not lobed; by sepals hairy on outside. Trees to 13 m tall by 1.4 m DBH (usually less than half these dimensions). Twigs densely covered with fine hairs when young, lacking thorns. Leaves rolled inward in the bud. Flowers about 3 cm across. Fruits ripening from early summer to early autumn; those of wild trees often small and sour and not of domestic quality, but can be used for jelly. Wood hard and used for crafts; excellent for firewood. Occasional. Cultivated and escaped; old fields, fencerows, thickets, abandoned homesites. To over 1500 m in sAppalachians. Mar–May. Syn: *Pyrus malus* L.; *Malus sylvestris* (L.) Mill.

2.
Prairie Crabapple;
Wild Crab
Malus ioensis (Wood)
Britt.
[250]

Recognized by leaves permanently hairy beneath, coarsely serrate, often lobed, folded in the bud; by sepals hairy on outside. Shrubs or trees to 10 m tall by 25 cm DBH. Stems sometimes with a few thorns. Flowers 3.5–5 cm across. Fruits 2.5–4 cm across, long-stalked, yellow-green, waxy, nearly globose, hard and bitter; can be used for jelly. The 5 spots on the leaf in photograph are apple-cedar rust, a common infection of apples. Sometimes planted as an ornamental, especially a double-flowered form, Bechtel's Crabapple. Occasional. Bottomlands, prairies, thin woods. To about 450 m elevation. Apr–May. Syn: *Pyrus ioensis* (Wood) Bailey.

3.
Southern Crabapple;
Wild Crabapple
Malus angustifolia (Ait.)
Michx.
[251]

Recognized by leaves on fertile twigs mostly crenate to crenate-serrate, rounded to obtuse to barely acute; by sepals glabrous on outside. Shrubs or trees to 14 m tall by 46 cm DBH, commonly sprouting from roots, frequently forming thickets. Plants uncommonly thorny; twigs often hairy when first formed, soon becoming glabrous. Leaves on vigorous twigs often lobed. Flowers quite fragrant, about 25 mm across. Fruits 20–35 mm across, long-stalked, yellowish-green, very sour, maturing in late summer; can be used for cider, jelly, and preserves. Sometimes planted as an ornamental, but is likely to become infected with apple-cedar rust if *Juniperus* trees are nearby. Occasional. Thin woods, roadsides, fencerows, old fields; most often in moist habitats. Feb–May. Syn: *Pyrus angustifolia* Ait.

4.
Sweet Crabapple;
Wild Crabapple
Malus coronaria (L.) Mill.
[252]

Recognized by mature leaves glabrous on underside
(may be quite hairy when young), those on fertile twigs
all serrate to doubly serrate, often some with triangular
lobes at widest part; by sepals glabrous on outside.
Shrubs or trees to 9 m tall by about 40 cm DBH.
Branches usually thorny; twigs with grayish hairs when
young, becoming glabrous. Leaves folded in the bud.
Flowers fragrant, 2–4 cm across. Fruits long-stalked,
nearly globose to somewhat flattened, 25–35 mm
across, yellowish-green, very sour, maturing in late
summer or early autumn; can be used for cider, jelly,
and preserves. Wood hard, used for crafts and firewood. Sometimes grown as an ornamental, but is
likely to become infected with apple-cedar rust if
Juniperus trees are nearby. Occasional. Fencerows, old
fields, woods margins; most frequently in moist soils.
To about 1000 m elevation in sAppalachians. Apr–
June. Syn: *Pyrus coronaria* L.

6.
CRATAEGUS. Hawthorn; Red Haw; Thorn

Recognized by having thorns, rarely absent; by fruits resembling small to tiny
apples, but with 1–5 bony and usually one-seeded stones. Shrubs or trees. Leaves
deciduous, 1 leaf and 1–2 buds at tip of vigorous twigs, end bud a terminal one.
Stipules and stipule scars present, the latter usually small and narrow. Flowers
bisexual, solitary or in clusters of a few to many. Sepals 5; petals 5, white or
sometimes pink; stamens 5–25; ovary inferior; styles 1–5, separate, persisting.
Fruits yellowish-green to yellow, orange, or red (bright blue and glaucous in *C.
brachycantha* Sarg. & Engelm., Blueberry Hawthorn); often persisting after leaves
fall; flesh firm to succulent. *Crataegus* is a large genus and the subject of much
controversy as to species definition and number. One manual, for example, includes
103 species for a given section of the US and another describes only 21 for the same
area. Possible reasons for this difficulty include hybridization and several unusual
aspects involving reproduction, namely: (1) *apomixis,* in which the embryo develops
from cells other than sex cells, (2) *polyploidy,* in which chromosome numbers may
increase by 50, 100, 150, 200 percent or more, and (3) *anauploidy,* in which
chromosome numbers may be other than multiples of the sex cells as found in
polyploidy. There are probably about 35 species in the SE, of which we are including a representative few to illustrate variation and promote easier recognition of the
genus. A key is not included. *Crataegus* spp. should not be ignored as they may be
expected in most types of habitats from river bottoms and wet depressions to sandy
scrub oak–pine woods and thin soils of rock outcrops. They are an important source
of food for wildlife, especially birds, through which viable seeds are widely distributed. *Crataegus* species are subject to rust infections, as are their similar relatives,
Malus. They occur in the N. Hemisphere, most abundant in c and neN. Amer.
Nearly 1000 species have been proposed.

May Haw
Crataegus aestivalis (Walt.)
T. & G.
[253]

Unusual among haws in having fruits ripening early, Apr–July. Recognized by leaves 3–5 cm long on flowering twigs, to 8 cm long on vigorous vegetative twigs, wedge-shaped at base, margin serrate to doubly serrate and rarely lobed, the latter usually only on vigorous twigs; by flowers solitary or in clusters of 2–3, sepals entire. Trees, less commonly shrubs, to 13 m tall by 40 cm DBH. Twigs glabrous, crooked; thorns 15–40 mm long, occasionally absent. Petioles absent or to 1 cm long. Fruits red, globose, 8–10 mm across; pulp soft, juicy, slightly acid; stones 3–5. Fruits are sold in markets locally for jellies that are tart and flavorful. Common. River swamps, pond margins, stream banks, wet woods. Feb–Apr.

Fanleaf Hawthorn
Crataegus flabellata (Bosc)
K. Koch
[254]

Variable, specimens not always identifiable with reasonable certainty. At least 15 species, as interpreted in some other treatments, are included in the species as described here. Recognized by leaves obtuse to rounded at base, petioles glandless or rarely with 1–2 glands near base of blade, lateral veins running only to points, not notches; by sepals entire or nearly so, rarely with 1–2 glands. Shrubs or trees to 9 m tall by 18 cm DBH. Thorns abundant to scarce, slender and usually curved. Leaves at first hairy, becoming glabrous, 4–8 cm long, serrate and often shallowly lobed, especially on nonflowering twigs. Flowers 13–20 cm across, in clusters of a few to many, petals white. Fruits maturing Sept–Oct, red, 8–15 mm across, stones usually 5. Common. Usually well-drained rocky areas; stream banks, pastures, old fields. To about 1800 m elevation in sAppalachians. May–June.

Southern Haw
Crataegus flava Ait.
[255]

Variable, specimens sometimes difficult to identify, described here to include perhaps a dozen or more species as treated by some authors. Recognized by leaves obovate, often shallowly lobed toward tip, margin serrate, the teeth tipped with tiny black glands, base of blade varying from tapered to wedge-shaped or rarely rounded; by petioles obscured by part of the blade extending down the edges and bearing black glands. Shrubs or trees to 7 m tall by 20 cm DBH. Young branches woolly; thorns to 6 cm long. Flowers about

17 mm wide, in clusters of 2–3, rarely solitary; pedicels hairy. Fruits maturing Aug–Sept, red, globose, 8–17 mm across; stones 5. *C. flava* as interpreted in some books has brown to yellow fruits. Common. Dry soil of rocky areas, well-drained sandy soils, oak-pine woods. Mar–May.

Parsley Hawthorn
Crataegus marshallii Eggl.
[256]

Easily recognized by its leaves, which are broadly ovate, 1–5 cm long, nearly as wide, deeply cut into lobes, the lobes coarsely toothed and sometimes bearing smaller lobes, the veins running both to notches and to points. Shrubs or trees to 10 m tall by 13 cm DBH. Branches slender; thorns to 3 cm long, slender, occasionally absent. Flowers on slim hairy stalks, to 15 mm across, in clusters of 3–12, petals white, anthers red. Fruits maturing Sept–Oct, oblong to obovoid, 5–9 mm long, bright red, flesh succulent, stones 1–3. Common. Bottomland and floodplain woods. Mar–May.

Spatulate Haw;
Littlehip Hawthorn
Crataegus spathulata
Michx.
[257]

Clearly distinguished from all other species by spatulate leaves with some 3-lobed and by lateral veins running both to notches and to points. Shrubs or trees to over 8 m tall by about 30 cm DBH. Bark gray, thin, peeling in patches and exposing brown areas. Thorns slender, to 4 cm long. Leaves 1–5 cm long, glabrous or hairy, blades usually unlobed on flowering twigs. Petioles obscured by blade tapering down the edges. Flowers 6–8 mm across, in many flowered clusters; petals white; anthers pale yellow. Fruits maturing Sept–Oct, red, nearly globose, 4–7 mm across, stones 3–5. Common. Usually in moist places, but also in well-drained upland areas; mostly in thin woods; stream banks, fencerows, pastures, along streams. Apr–May.

Dwarf Haw; One-flower
Hawthorn
Crataegus uniflora
Muenchh.
[258]

This *Crataegus* is fairly easy to recognize if flowers or fruits are present. Identified by leaves serrate, the teeth with tiny black glands on tips, lower part of blade lacking teeth and glands, veins running only to points of lobes, base wedge-shaped, petioles absent or to 5 mm long; by flowers 10–15 mm across, solitary or rarely in clusters of 2–3; by sepals coarsely glandular-serrate, persistent and conspicuous on fruits. Low shrubs or rarely trees to 5 m tall by 15 cm DBH.

Twigs and branches usually crooked, thorns slender. Fruits maturing Sept–Nov, nearly globose, 9–13 mm across, brownish to reddish; flesh dry to mealy; stones 5 or uncommonly 3–4. Common. Usually dry soils; rocky and sandy places; thin woods, old fields. Mar–May.

7.
PRUNUS. Peach; Cherry; Plum

Recognized by a bitter-almond odor usually detectable in fresh twigs when broken; by lenticels horizontally elongated on stems over 5 mm across, bud scales 4–6, stipules present but soon falling; by fruits drupes with one stone. Shrubs or trees, sometimes with thorns. Bark often exuding a gummy substance (a polysaccharide), especially if damaged; outer layer of bark peeling in horizontal strips in most species. Buds sessile, axillary buds solitary or 2–3 side by side; bundle scars 3; stipule scars small and often indistinct. Leaves deciduous or evergreen. Flowers bisexual, showy, solitary or in clusters. Flowers with a cuplike structure (hypanthium) bearing on its rim 5 sepals, 5 petals, and numerous stamens; bearing in the bottom a single pistil containing 2 ovules. The hypanthium usually falls off or is ruptured by the enlarging fruit. The stone in the fruits contains one seed, rarely 2. Various species are cultivated for food (Peach, Apricot, Nectarine, Plum, Cherry, Almond) and as ornamentals (Japanese Cherry, Cherry-almond, Cherry-plum, Laurel Cherry). Fruits of some species in the wild provide food for man and can be used for jams, jellies, preserves, and alcoholic beverages. Fruits of all species are eaten abundantly by wildlife with resulting profuse and widespread dissemination as the seed remains viable in stones dropped in feces. There are more than 400 species in the genus, mostly in temperate regions, all in the N. Hemisphere and the Andes of S. Amer. Tree species native to the US number about 18, with 12 occurring in the SE. Some cultivated introductions have escaped and reproduce naturally. Identification to species can be troublesome as inspection of minute characteristics may be needed, especially if flowers and fruits are not available. Reference to distribution may be helpful.

KEY TO PRUNUS SPECIES

1. End bud terminal one (Peach, cherries)   2
    2. Buds with gray to tan woolly hairs, flowers and fruits sessile or nearly so
                                   1. *P. persica*
    2. Buds lacking woolly hairs, flowers and fruits on conspicuous stalks   3
      3. Leaves semi-evergreen, petioles nonglandular; flesh of fruits somewhat mealy   2. *P. caroliniana*
      3. Leaves deciduous, petioles usually glandular near the blade; flesh of fruits juicy   4
        4. Twigs finely hairy, often nearly glabrous late in season; leaves more than half as wide as long   4. *P. mahaleb*
        4. Twigs glabrous, sometimes hairy when young; leaves about 2 × or more as long as wide   5

     5. Flowers and fruits in elongated clusters of about 20 or more   6

       6. Leaves with sharp spreading teeth, sepals blunt   3. *P. virginiana*

       6. Leaves with blunt incurved callus-tipped teeth, sepals acute

          5. *P. serotina*

     5. Flowers and fruits solitary or in rounded or flat-topped clusters of 12 or less   7

       7. Leaves lanceolate, long-acuminate   6. *P. pensylvanica*

       7. Leaves ovate-elliptic or ovate-oblong to obovate, acute to acuminate   8

         8. Leaves 8–15 cm long, hairy beneath, with 10–14 prominent veins on each side of midrib, glands on distal end of petiole (occasionally 1–2 on base of blade); fruits sweet   7. *P. avium*

         8. Leaves 5–9 cm long, glabrous, with 6–8 prominent veins on each side of midrib, glands usually present on margin at base of blade, none on petiole; fruits sour   8. *P. cerasus*

1. End bud an axillary one (plums)   9

   9. Leaves with deep purplish-red color masking most of the green, buds hairy and acute   9. *P. cerasifera*

   9. Leaves green except for autumn color changes, buds seldom both hairy and acute   10

     10. Teeth of leaves with a gland, this sometimes falling with age leaving a scar (callus)   11

       11. Leaves mostly under 2 cm wide, teeth at mid-leaf about 20 per cm; flowers under 10 mm across   10. *P. angustifolia*

       11. Leaves mostly over 2 cm wide, teeth at mid-leaf 8–15 per cm; flowers 12 mm or more across   12

         12. Flowers opening before or at beginning of leaf expansion; teeth of leaves with the gland below the tip   11. *P. munsoniana*

         12. Flowers opening as or after leaves expand; teeth of leaves with the gland on very tip   12. *P. hortulana*

     10. Teeth of leaves glandless   13

       13. Leaves obtuse to acute at apex, velvety hairy beneath   13. *P. maritima*

       13. Leaves acute to acuminate at apex, always some leaves short-acuminate, glabrous or hairy beneath   14

         14. Petioles with 2 or more (rarely 1) glands near blade, blades hairy beneath   14. *P. mexicana*

         14. Petioles usually without glands near blade, blades glabrous or hairy beneath   15

           15. Leaves ordinarily doubly serrate; flowers 18–30 mm across; fruits 18–30 mm across, yellow to red, stone over 10 mm long   5. *P. americana*

           15. Leaves crenate to serrate; flowers 8–15 mm across; fruits 10–15 mm across, dark red to purple or nearly black; stone 8–12 mm long   16

             16. Plants of the Piedmont and CP of the SE   16. *P. umbellata*

             16. Plants of the mountains of neTenn to Pa   17. *P. alleghaniensis*

1.

Peach

Prunus persica (L.) Batsch

[259]

Recognized by end bud a terminal one, buds with gray to tan woolly hairs; by flowers and fruits sessile or nearly so. Trees to 10 m tall by 35 cm DBH in cultivation; DBH to about 20 cm in the wild. Bark on young trunks smooth, older trunks with scaly plates that are marked by remains of the horizontally elongated lenticels. Twigs smooth, glabrous, shiny, green to red, when red usually with some green on the shaded side. Leaves serrate, the teeth with a tiny gland or callus; larger glands on distal end of the petioles and/or near base of the blades, or absent on some leaves. Flowers appearing before the leaves, about 3 cm across or smaller or to 5 cm across in some horticultural varieties, developing from 1–2 flower buds at the side(s) of axillary vegetative buds. Petals pink, or rarely whte. Fruits densely and finely hairy, variable in size and shape, usually 2–8 cm across, but larger in some horticultural varieties. Stone of fruits pitted and wrinkled, easily freed from the flesh (freestones) or the flesh adhering (clings). Ornamentals include dwarf, double-flowered, and purple-leaved varieties. Rare. Persisting and escaping in waste places, roadsides, trash dumps, edge of woods. A native of China; originally thought to come from Persia, hence the name, *P. persica.* Indians are believed to have introduced the species in numerous localities soon after it was brought to N. Amer by Europeans. Persisting to 1100 m elevation in sAppalachians. Feb–Apr.

2.

Carolina Laurel Cherry; Laurelcherry

Prunus caroliniana (Mill.)

[260]

Recognized by end bud a terminal one; by leaves semi-evergreen, petioles lacking glands; by flesh of fruits somewhat mealy. Trees to 14 m tall by 1 m DBH. Bark dark gray, tight and smooth, with prominent horizontally elongated lenticels on younger trunks, older trunks with shallow vertical and horizontal fissures breaking the bark into thin squarish plates. Twigs glabrous. Leaves 5–12 cm long, shiny above, entire or with short sharp teeth that are mostly wide-spaced. Flowers about 5 mm across, nearly white, in short compact racemes from the axils of leaves of the previous year. Fruits dull black, 10–13 mm long, flesh thin, the stone mostly ovoid and to 12 mm long. Some fruits may persist until after flowers appear the following year. Widely used as an ornamental, for hedges, and for screening. Fruits eaten by birds that widely distribute the stones with their viable seed. Leaves, stems, and seeds poisonous, producing hydrocyanic acid (prussic acid) when wilted or eaten. Common. Thin woods, maritime woods, fencerows, vacant lots, fields; often forming dense thickets. Feb–Apr.

3.
Common Chokecherry
Prunus virginiana L.
[261]

Recognized by glabrous twigs, end bud a terminal one; by leaves deciduous, about 2 × as long as wide, with sharp spreading teeth. Shrubs or trees to 20 m tall by 52 cm DBH. Bark reddish-brown to grayish-brown, at first smooth and with horizontal lenticels, becoming shallowly fissured with age. Broken twigs and inner bark with a strong rank odor and a bitter taste. Buds glabrous, the largest ones 6–8 mm long. Leaves smooth above, petioles usually with 2 glands near base of blade. Flowers 8–12 mm across, in racemes that are terminal on leafy twigs of the season; sepals blunt; hypanthium and other attached flower parts withering and falling as fruit develops. Fruits 6–10 mm across, red to blackish, the flesh juicy and astringent. Raw seeds poisonous, rendered harmless by cooking; foliage and stems, especially the bark, also poisonous when eaten. Widely and abundantly spread by birds and other animals that eat the fruits. Much like *P. serotina,* which may be distinguished by the blunt incurved teeth on leaf margins. Common. Borders of woods, thin woods, roadsides, fencerows, along streams, dunes, rocky areas. Extends westward to Cal, and north to swNWT, Ont, and Nfld. Apr–June.

4.
Mahaleb;
Perfumed Cherry
Prunus mahaleb L.

Easily recognized by twigs finely hairy, often nearly glabrous late in the season, and by leaves distinctly more than ½ as wide as long. Trees to 10 m tall by 60 cm DBH in cultivation. Trunk short and often crooked; bark and wood aromatic-fragrant; twigs pale green. Leaf margin finely crenate-serrate with tiny glands between teeth. Flowers fragrant, 10–16 mm across, in green leafy-bracted clusters of 4–10. Fruits 6–10 mm across, dark red to black, pulp bitter, un-palatable. Planted as an ornamental and escaping. Rare. Roadsides, fencerows, woods, borders. Native of Eurasia. Apr–May.

5.
Black Cherry;
Wild Cherry
Prunus serotina Ehrh.
[262]

Recognized by glabrous twigs, end bud a terminal one;
by leaves deciduous, over 2 × as long as wide, rarely
less, with blunt incurved teeth. Trees to 35 m tall by
2.3 m DBH. Bark dark reddish-brown to nearly black,
smooth and with horizontal lenticels, becoming fissured and scaly with age; inner bark and broken twigs
with an almondlike odor and bitter taste. Buds
glabrous, the largest about 4 mm long. Leaves smooth
above, base of blade and adjacent part of petiole with 1
or more glands. Flowers 7–10 mm across, in racemes
that are terminal on leafy twigs of the season, petals
white, sepals acute. Hypanthium and sepals persisting
below fruit. Fruits 7–10 mm across, dark red to almost
black, the flesh juicy, usually sweet, but sometimes
slightly bitter. A poison, prussic acid, is released from
vegetative parts and seeds when eaten raw; however,
cooking dissipates the poison. Probably more livestock
are killed from eating Black Cherry than from any
other plant. The wood is close-grained, hard, and
highly prized for furniture, paneling, interior trim,
veneers, crafts, and other uses. Two subspecies are generally recognized in the SE. The more abundant one,
serotina, has leaves glabrous beneath except for perhaps
some hairs on the veins, and the axis of the racemes is
glabrous. The other is ssp. *hirsuta* (Ell.) McVaugh
which has leaves abundantly hairy beneath and the axis
of the racemes hairy. This subspecies has also been
treated as a variety, var. *alabamensis* (Mohr) Little, and
includes *P. cuthbertii* Small. It is of limited distribution
in scattered localities in NC and SC, and from eGa to
neAla and nwFla. *P. serotina* is much like *P. virginiana,*
which may be distinguished by its sharp spreading
teeth on leaf margin. Common. Widely and abundantly spread by birds and other animals that eat the
fruits. Forests, old fields, fencerows, pastures. Grows
to over 1530 m elevation in sAppalachians. Occurs
northward to NS, west to cAriz, south into Mex and
Guatemala. Mar–June.

6.
Pin Cherry; Fire Cherry
Prunus pensylvanica L. f.
[263]

Recognized by end bud a terminal one; by leaves deciduous, lanceolate, long-acuminate; by flowers and fruits
solitary or in rounded or flat-topped clusters of 12 or
less. Shrubs or trees to 27 m tall by 55 cm DBH. Bark
smooth, reddish-gray, thin, with age fissured into scaly
plates, the outer layer easily peeled. Twigs glossy, lenticels and pith orange. Leaves thin, wrinkled, 6–15 cm

long, glossy on both sides, finely serrate, the teeth red and callus-tipped. Flowers 12–17 mm across, in umbels of 2–11, rarely solitary, subtended only by bud scales; sepals glabrous. Fruits 5–8 mm across, bright red, sour, with thin juicy flesh; stone slightly flattened. Wood soft and porous, with minor value for pulp and fuel. Fruits an important source of food for birds and various other wildlife species that spread the Cherry widely and abundantly. Important as reforesting agent after fires and lumbering, providing cover and shade for establishment of other tree species. Pin Cherry usually begins dying in about 25 years, being replaced by more valuable timber trees. Common. Burned and lumbered areas, other openings. To nearly 2000 m elevation in sAppalachians. Occurs in scattered localities north to Nfld and west to Col and BC. Mar–June.

7.
Sweet Cherry;
Mazzard Cherry
Prunus avium (L.) L.
[264]

Recognized by glabrous twigs, end bud a terminal one; by leaves deciduous, petiole usually glandular near blade; by flowers and fruits stalked, solitary or in umbels from twigs of previous year, subtended only by bud scales. Trees to 21 m tall by 1.9 m DBH in cultivation. Bark reddish-brown, glossy, becoming dark, thick, and furrowed with age. Leaves 8–15 cm long, serrate to doubly serrate, the teeth slightly rounded, with 10–14 prominent veins on each side of midvein. Flowers 25–30 mm across; petals white, opening before or as leaves unfold. Fruits 10–25 mm across, red to purplish, flesh firm and sweet, stone usually ellipsoid. Rare. Escaping to roadsides, fencerows, woods margins, thin woods. Native of Eurasia. Apr–May.

8.
Sour Cherry
Prunus cerasus L.

Similar to Sweet Cherry but may be recognized by leaves 5–9 cm long, with 6–8 prominent veins on each side of midrib, and glands usually at base of blade. Trees to 20 m tall by 95 cm DBH in cultivation, often suckering from roots. Bark reddish-brown, glossy, becoming scaly and almost black with age. Flowers 15–25 mm across. Commonly cultivated for the fruits, more often in the cooler parts of the SE and northward. Rare. Escaping to fencerows, fields, roadsides, woods margins. Native of Asia. Apr–May.

9.
Cherry Plum;
Myrobalan Plum
Prunus cerasifera Ehrh.
[265]

Recognized by leaves with deep purplish color that masks most of the green color; by buds hairy and acute. Trees to 7 m tall by 35 cm DBH, frequently suckering from roots, spreading in this manner as well as by seeds. Twigs glabrous, slender, end bud an axillary one. Leaves thin, to about 6 cm long, finely serrate. Flowers about 25 mm across, solitary or in clusters of 2–3 from twig of previous year, petals pink. Fruits to about 25 mm long, juicy and sweet. There are several horticultural varieties of *P. cerasifera*, the only one we have found reproducing naturally being var. *atropurpurea* Dipp., which is described here. Other varieties, some with double flowers, generally have green leaves. The purplish-leaved variety is commonly planted as an ornamental. Rare. Escaping to vacant areas, fencerows, pastures. Native of swAsia. Apr–May.

10.
Chickasaw Plum
Prunus angustifolia Marsh.
[266]

Recognized by end bud an axillary one, buds about as wide as long; by leaves under 2 cm wide, teeth at midleaf about 20 per cm, each tooth with a tiny reddish gland which may fall with age leaving a callus. Shrubs or trees to about 6 m tall by over 50 cm DBH; often forming thickets, largely by means of root suckers. Twigs slender, slightly zigzag, short lateral twigs often ending as a thorn. Leaves frequently clustered on short spurs, blades tending to fold upward lengthwise making them troughlike. Flowers 6–9 mm across, solitary or in clusters of 2–4 from twigs of previous year, opening before leaves appear; petals white. Sepals without glands, hairy on upper side, glabrous below, falling with the hypanthium as the fruit enlarges. Fruits ripening in early summer, 9–20 mm across, yellow to red, largely spherical but variable in size, shape, and quality; flesh thick, juicy, edible although sometimes sour, used for jellies and preserves. Fruits an important source of food for wildlife, which spread the species widely and profusely. Common. Roadsides, fencerows, pastures, old fields, woods borders. Extends to neNMex and seCol. Possibly spread beyond original range by Indians. Feb–Apr.

11.
Wild Goose Plum
Prunus munsoniana
Wight & Hedr.

Recognized by end bud an axillary one; by leaves over 2 cm wide, teeth at midleaf 8–15 per cm, each tooth with a gland below the tip; by flowers opening before or at beginning of leaf expansion. Trees to 9 m tall by 38 cm DBH. Twigs glabrous. Leaves 6–10 cm long, longitudinally rolled in bud, when mature tending to fold upward lengthwise in a troughlike fashion. Flowers 12–16 mm across, solitary or in clusters of 2– 5 from twigs of previous year; petals white; sepals with marginal glands. Fruits red, rarely yellow, 12–20 mm across; flesh firm, juicy, sweet, tart, ripening in early summer, used for jelly and preserves; eaten by wildlife. Occasional. Stream banks, floodplains, rich woods, pastures. Apr.

12.
Hortulana Plum
Prunus hortulana Bailey

Similar to *P. munsoniana* but may be distinguished by teeth of leaves with a gland on the tip; by flowers opening as or after leaves expand. Trees to over 8 m tall by 27 cm DBH. Stems sometimes thorny. Leaves folded longitudinally in bud, flat at maturity. Fruit 2– 3 cm across, ripening in late summer or early fall. Occasional. Similar habitats. Apr.

13.
Beach Plum
Prunus maritima Marsh.
[267]

Recognized by end bud an axillary one; by leaves velvety hairy beneath, the tip obtuse to barely acute, teeth glandless. Shrubs, rarely trees to 5 m tall. Twigs hairy when young, nearly glabrous toward end of summer; buds hairy. Leaves firm; dull, dark green and glabrous above, paler beneath; 3–7 cm long, about half as wide; finely serrate, the teeth rounded but sharp-pointed. Flowers 12–20 mm across, solitary or in clusters of 2–3, on last year's twigs, appearing before the leaves; pedicels hairy, subtended only by bud scales. Fruits nearly globose, purplish-black, rarely red or yellow, 13–25 mm across, ripening Sept–Oct; eaten by wildlife, used for preserves and jelly. Occasional. Dunes, sandy soils near coast. Md to NB. Apr–June.

14.
Mexican Plum;
Big-tree Plum
Prunus mexicana S. Wats.
[268]

Recognized by end bud an axillary one; by leaves green, abruptly acuminate, hairy beneath, teeth sharp and glandless, petioles with glands near base of blade. Trees to 12 m tall by 45 cm DBH, plants solitary, having no root suckers. Twigs glabrous, shiny. Leaves 5–12 cm long, serrate to doubly-serrate, upper side wrinkled due to conspicuously sunken veins. Flowers 15–20 mm across, opening before leaves, solitary or in clusters of 2–4 on last year's twigs; petals white. Fruits ripening in late summer, dark purplish-red, glaucous, nearly globose, 20–25 mm across, pulp sweet; eaten by wildlife and man, used for jelly and preserves. Common. Moist to dry habitats; thin woods, prairies, river bottomlands, lake shores. Extends into neMex. Mar–Apr.

15.
American Plum
Prunus americana Marsh.
[269]

Recognized by end bud an axillary one; by leaves usually doubly serrate, teeth glandless, petioles without glands near blade; by flowers 18–30 mm across; by fruit 18–30 mm across. Shrubs or trees to 11 m tall by 30 cm DBH, with root suckers. Short lateral twigs often ending as a thorn. Leaf tips acuminate. Flowers fragrant, with white petals, opening before or as leaves develop, solitary or 2–4 in clusters on twigs of previous year. Fruits red, rarely to yellow, glaucous, ripening in late summer; flesh yellowish, juicy, sour. Fruits eaten fresh and used for jelly, preserves, cakes, puddings; eaten by wildlife, including deer, bears, raccoons, squirrels, and birds. Plants with petioles, lower leaf surface, and sepals persistently soft hairy treated by some authorities as *P. americana* var. *lantana* Sudw. Several cultivated varieties with improved fruit or appearance have been developed. Common. Usually moist habitats; sandy terraces along streams, thin woods, fencerows, pastures, abandoned fields. To over 1100 m elevation in sAppalachians. In scattered localities west to cwMont and cNMex. Feb–May.

16.
Hog Plum;
Flatwoods Plum
Prunus umbellata Ell.
[270]

Recognized by end bud an axillary one; by leaves acute to acuminate, margin serrate, teeth glandless, most or all petioles without glands near blade; by flowers 9–15 mm across; by fruits 10–15 mm across, pedicels 14–20 mm long. Shrubs or trees to 10 m tall by 32 cm DBH, sometimes in small colonies, but rarely with root suckers. Trunk often crooked. Leaves crenate to crenate-serrate, teeth 0.3–0.8 mm high, the base of blade usually with 2 dark glands, rarely some petioles with 1 or 2 glands at distal end; blades glabrous or hairy on lower or both surfaces, petioles glabrous or hairy. Flowers with little fragrance, solitary or in clusters of 2–5 on last year's twigs, opening before leaves appear; petals white. Fruits very dark purple, sometimes varying to red or yellow, very glaucous; flesh thick, sour, and occasionally also bitter; ripening July–Sept. Hog Plum varies considerably and has been divided into several species by some authors and into varieties by others, based mainly on hairiness. Under such a division *P. mitis* Beadle has twigs glabrous and leaves hairy on both surfaces, *P. injucunda* Small has twigs and both surfaces of leaf hairy, and *P. umbellata* has twigs and leaves glabrous or leaves hairy only on lower surface. Much like *P. alleghaniensis* and apparently separated only by their different distributions. Common. Stream banks, thin woods, dunes, flatwoods, rocky areas. Feb–Apr.

17.
Allegheny Plum
Prunus alleghaniensis
Porter

Having the same characteristics as some of the variants of *P. umbellata* and apparently may be distinguished only by their distributions. Shrubs or uncommonly trees to 6 m tall by 33 cm DBH in cultivation, thicket-forming. Sometimes thorny. Fruiting Aug–Sept. Rare. Mountain slopes, thin woods, rocky places. Apr–May.

AQUIFOLIACEAE: Holly Family

8.
ILEX. Holly

Six species of *Ilex* reaching tree size in the SE are evergreen and are treated in Group
H. Six species are deciduous and may be recognized by year-old twigs not green,
buds sessile, the end bud a terminal one, bundle scar 1, pith continuous, fresh twigs
without a bitter-almond odor when crushed; by tiny dark sharp-pointed stipules (the
stipules best seen at junction of current and previous year's twigs, both terminal and
lateral ones); by 1 leaf at end of vigorous twigs, leaf blades longer than wide. In
hollies of the SE flowers are small, solitary or in clusters from leaf axils, usually of
one sex, the sexes on separate plants, or rarely on the same. Petals white to greenish-
white, 4 or uncommonly 5–6. Stamens of same number as petals, sometimes
present and sterile in fertile flowers. Fruits drupes, with bitter pulp in most species,
in deciduous species with 2–10 1-seeded stones (nutlets); an important source of
food for wildlife. Limited use as an ornamental is made of deciduous hollies bearing
fruit. Identification to species is often a problem and may require the presence of
flowers and/or fruits. Even with these present, certain plants may be difficult to
name with confidence, and without them identification may be impossible. A genus
of over 300 species of shrubs and trees in N. and S. Amer, Asia, and a few in Africa.
All species reaching tree size in the US occur east of the Rocky Mts.

KEY TO DECIDUOUS ILEX SPECIES
1. Leaves with crenate margin   2
   2. Fruiting pedicels 2–8 mm long   1. *I. decidua*
   2. Fruiting pedicels 12–20 mm long   2. *I. longipes*
1. Leaves with serrate margin, sometimes entire, rarely crenate-serrate   3
   3. Plants with fruits   4
      4. Nutlets smooth on back   3. *I. verticellata*
      4. Nutlets grooved or ribbed on back   5
         5. Leaves entire to finely serrate, quite wrinkled above, sepals not ciliate
                                                                         4. *I. amelanchier*
         5. Leaves serrate to crenate-serrate, usually faintly wrinkled above, sepals
            ciliate   6
            6. Leaves 3–8 cm long, marginal teeth usually inconspicuous; fruits 5–
               9 mm across   5. *I. ambigua*
            6. Leaves 6–16 cm long, marginal teeth prominent; fruits 9–12 mm
               across   6. *I. montana*
   3. Plants with flowers   7
      7. Plants with male flowers   8
         8. Flowers 3 or more in peduncled clusters   9
            9. Sepals ciliate   3. *I. verticellata*
            9. Sepals not ciliate   4. *I. amelanchier*
         8. Flowers pedicellate only, not peduncled   10

10. Leaves 3–8 cm long, marginal teeth usually inconspicuous
                                                                 5. *I. ambigua*
    10. Leaves 6–16 cm long, marginal teeth prominent   6. *I. montana*
7. Plants with female flowers   11
   11. Pedicels 7 mm or more long, sepals not ciliate   4. *I. amelanchier*
   11. Pedicels under 7 mm long, sepals ciliate   12
     12. Flowers mostly in axils of leaves on elongated twigs
                                              3. *I. verticellata*
     12. Flowers mostly on short lateral spurs   13
       13. Leaves 3–8 cm long, marginal teeth usually inconspicuous
                                              5. *I. ambigua*
      13. Leaves 6–16 cm long, marginal teeth prominent
                                            6. *I. montana*

1.
Possum-haw
Ilex decidua Walt.
[271]

Recognized by crenate leaf margin; by fruiting pedicels 2–8 mm long. Shrubs or trees to 10 m tall by 20 cm DBH. Terminal bud obtuse. Leaves usually widest at or beyond the middle, 2.5–7.5 cm long. Flowers on pedicels 2–8 mm long, usually on short lateral spurs; sepals, petals, and stamens 4 each; sepals not ciliate. Fruits red or rarely yellow, 4–8 mm across, globose or nearly so, pulp bitter, nutlets 4.5–5 mm long and irregularly grooved on the back. Some plants in nFla with leaves only 1–2 cm long, have been named *I. decidua* var. *curtissii* Fern. Occasional plants of *I. ambigua* have somewhat crenate leaf margins and may be confused with *I. decidua,* but may be distinguished by having sepals, petals, and stamens 5 each. Common. Floodplains, swamps, margins of streams, lakes, and ponds; occasionally in upland areas. To about 360 m elevation. Mar–May.

2.
Georgia Holly
Ilex longipes Chapm. ex Trel.

Much like Possum-haw, but has flowering and fruiting pedicels 12 mm or more long, with fruiting pedicels as long as 30 mm. With flowers and fruits lacking, the two species are probably not separable. Georgia Holly is considered by some a variety of Possum-haw. Occasional. Rocky slopes, upland woods. Mar–May.

3.
Common Winterberry;
Black-alder
Ilex verticellata (L.) Grav
[272]

Recognized by leaf margin serrate, rarely crenate-serrate; by sepals ciliate; by nutlets of fruit 3–4 mm long and smooth on the back. Shrubs or rarely trees to 8 m tall by 10 cm DBH. Leaves 4–10 cm long, usually hairy beneath, the veins prominent. Staminate flowers in peduncled clusters; pistillate flowers and fruits with pedicels only. Fruits red, 5–7 mm across, globose, ordinarily with 5 or 10 nutlets. If flowers and fruits are lacking, plants of this species and *I. montana* may be difficult to separate. Generally leaves of the former are firm and the larger veins are conspicuously sunken on the upper surface, while those of the latter are membranous and the veins are not conspicuously sunken. Occasional; rare as a tree. Swamps, bogs, along streams, wet woods. Occurs northward to ceNfld. Apr–June.

4.
Sarvis Holly
Ilex amelanchier
M. A. Curtis
[273]

Recognized by leaf margin entire to shallowly serrate, the base obtuse to rounded, upper surface wrinkled; by sepals not ciliate; by nutlets 4, with 2 deep furrows on the back. Shrubs or rarely trees to 5 m tall by 8 cm DBH. Staminate flowers 3–several on axillary peduncles; pistillate flowers solitary in left axils. Fruits red, 8–10 mm across, globose or nearly so. Rare, but common locally. Moist to wet places; swamps, stream banks, pond margins, clay-based bays. To about 60 m elevation. Apr–May.

5.
Carolina Holly
Ilex ambigua (Michx.)
Torr.
[274]

Recognized by leaves 3–8 cm long, margin finely to coarsely serrate or rarely crenate-serrate, the teeth usually inconspicuous; by flowers pedicellate only, the sepals ciliate; by fruits 5–9 mm across, the nutlets 4–7 mm long and conspicuously furrowed on the back. Shrubs or rarely trees to 6 m tall by 13 cm DBH. Flowers and fruits on short pedicels. Fruits red, rarely yellow, globose to ellipsoid. Plants with leaves over 5 cm long and fruits over 8 mm across are similar to some individuals of *I. montana*. The latter ordinarily can be recognized by having prominent marginal teeth. Occasional. Upland woods, commonly in sandy soils. To about 300 m elevation. Mar–June. Syn: *I. caroliniana* (Walt.) Trel.

6.
Mountain Winterberry
Ilex montana T. & G.
[275]

Identified by leaves 6–16 cm long, margin sharply serrate; by flowers pedicellate only, the sepals ciliate; by fruits 9–12 mm across, the 4 nutlets furrowed on the back. Shrubs or uncommonly trees to 12 m tall by 20 cm DBH. Terminal buds pointed. Flowers and fruits on short pedicels. Fruits red, globose. Some plants are similar to individuals of *I. ambigua* which see for separation. Mountain Winterberry is sometimes classified as a variety of Carolina Holly under the name *I. ambigua* var. *monticola* (Gray) Wunderlin & Poppleton or *I. ambigua* var. *montana* (T. & G.) Ahles. Common. Moist woods, mountain slopes, occasionally in dry thin woods. To over 1800 m elevation in sAppalachians. Apr–July. Syn: *I. montana* var. *mollis* (Gray) Britt.; *I. montana* var. *beadlei* (Ashe) Fern.

RHAMNACEAE: Buckthorn Family

9.
RHAMNUS. Buckthorn

Carolina Buckthorn;
Polecat-tree
Rhamnus caroliniana Walt.
[276]

Recognized by twigs lacking thorns, without bitter-almond odor when broken, end bud a terminal one, lateral buds sessile, pith continuous, stipules or their scars present; by one leaf at end of vigorous twigs, leaves longer than wide, lateral veins on each side of midvein evenly spaced; by fruits drupes with 1–4 seed-like stones. Shrubs or trees to 14 m tall by 33 cm DBH. Twigs angled. Mature buds hairy, tiny, lacking scales. Leaves 5–13 cm long, the margin minutely crenate-serrate, the tip acute, veins conspicuous on underside, with odor when crushed somewhat like that of a polecat (skunk). Flowers mostly in peduncled umbels; bisexual; about 5 mm across; sepals, petals, and stamens 5 each. Fruits turning red then black, pulp sweet; used as food by wildlife. Occasionally planted as an ornamental. *R. lanceolata* Pursh, which occurs in the SE from WVa to Ala and westward, is reported as reaching tree size only outside the SE. Should this species be encountered, it may be distinguished from *R. caroliniana* by leaves with acuminate tips and inconspicuous veins on underside; by sepals, petals, and stamens 4 each. Occasional. Usually in moist places; deciduous woods, especially in areas of limestone rocks; stream valleys. To about 600 m elevation. Extends into neMex. Four other native species reaching tree size occur in wUS. Apr–June.

THEACEAE: Tea Family

10.
FRANKLINIA. Franklinia

Franklinia
Franklinia alatamaha
Bartr. ex Marsh.
[277]

Recognized by twigs dark reddish-brown, buds with tan hairs, end bud a terminal one, lateral buds sessile, bundle scar 1 and not conspicuously protruding, stipules absent, pith continuous. Shrubs or trees to 7 m tall by 26 cm DBH. Twigs slightly angled. Leaves 6–18 cm long, turning orange or red in autumn. Fruits globose, 15–20 cm across, maturing in autumn, splitting lengthwise both from tip to middle and from base to middle. Seeds several, flat, wingless. Last seen in the wild in 1790 near Fort Barrington, Ga, on the ne side of the Altamaha R. close to the coast. Now apparently extinct and known only in cultivation. Related to *Gordonia lasianthus,* which is evergreen and differs in other ways such as having pointed fruits. July–Aug.

ERICACEAE: Heath Family

11.
CLETHRA. Clethra

Sweet Pepperbush;
White-alder
Clethra acuminata Michx.
[278]

Recognized by end bud on twigs a terminal one, lateral buds sessile, bundle scar 1 and conspicuously protruding from leaf scar; by leaves serrate. Shrubs or rarely trees to 6 m tall by 13 cm DBH. Bark reddish-brown, thin. Leaves 2–20 cm long, acuminate at tip. Petals 5–7 mm long, filaments hairy. Fruits nearly globose, hairy, 4–6 mm long, splitting into 3 sections. Sometimes planted as an ornamental. Another species of *Clethra* occurs in the SE, *C. alnifolia* L., but it is a shrub to 2.5 m tall. *Clethra* spp. are sometimes separated from the Heath Family and placed in the White-alder Family, the Clethraceae. Common. Rich woods of slopes and coves. To over 1650 m elevation in sAppalachians. *Clethra* is a genus of around 100 species in the Americas, Europe, and Asia. June–Aug.

12.
OXYDENDRUM. Sourwood

Sourwood
Oxydendrum arboreum (L.)
DC.
[279]

Recognized by end bud an axillary one, lateral buds sessile and solitary in each leaf axil, bundle scar 1, stipules absent; by leaf margin finely serrate. Shrubs or trees to 35 m tall by 73 cm DBH with a slender crown and branches often drooping toward ends. Twigs and leaves with a sour-tasting sap. Axillary buds partly embedded in bark. Leaves turning brilliant red in autumn. Flower and fruit clusters pendulous, borne terminally and thus prominent on outer fringe of the crown. Fruits narrowly ovoid 5-carpelled capsules 5–7 mm long, erect on a curved pedicel, generally persistent through winter. Seeds tiny, many. Wood hard, occasionally used in paneling, for tool handles, and in crafts; sourwood honey highly prized; planted as an ornamental. Common. Trees generally scattered; in well-drained woods, on bluffs, in secondary growth. To over 1700 m elevation in sAppalachians. Sourwood is the only species in the genus. May–July.

STYRACACEAE: Storax Family

13.
HALESIA. Silverbell

Recognized by pith chambered. Shrubs or trees. Buds with 4 visible scales, terminal bud an axillary one. Axillary buds often superposed; the lowest one may be quite small or even invisible without magnification; however, they are usually readily evident on vigorous twigs. Flowers and fruits dangling; flowers in clusters of 2–6, rarely 1, developing from axillary buds on twigs of previous year. Petals 4, united; sepals 4, persistent on fruit; stamens 8–16; ovary inferior. Fruits dry, winged, indehiscent, 1-seeded (rarely 2–3-seeded), the style persistent. Vegetatively some individuals of *Halesia* are similar to some *Styrax*. They may be separated by the 2 uppermost superposed buds of the former being contiguous and triangular whereas those of the latter are well separated and thumblike in shape. *Halesia* is a genus of 4 species, 3 in eUS and 1 in China. Because of recent changes in scientific names of 2 US species, one should be alerted to the possibility of confusion. Attention to the common names may be helpful, a reversal of the usual roles played by common and scientific names.

KEY TO HALESIA SPECIES

1. Leaves broadly obovate to suborbicular; corolla lobes longer than the corolla tube, petals united only near their bases; fruits with 2 lengthwise wings   1. *H. diptera*
1. Leaves elliptic, ovate, to oblong; corolla lobes shorter than corolla tube; fruits with 4 lengthwise wings   2
   2. Corollas over 13 mm long; fruits 3–5 cm long, ellipsoid to obovoid
                                                                     2. *H. tetraptera*
   2. Corollas under 13 mm long; fruits under 3 cm long, tapering toward the base
                                                                     3. *H. carolina*

1.
Two-winged Silverbell
Halesia diptera Ellis
[280]

Recognized by leaves obovate to nearly orbicular. Shrubs or trees to 17 m tall by 40 cm DBH. Leaves 6–12 cm long, 4–10 cm wide, finely and irregularly serrate. Corollas 10–32 mm long, deeply 4-lobed, the petals sometimes appearing separate. Fruits 3–5 cm long, with 2 prominent lengthwise wings; unripe fruits, which are sour, eaten by wildlife. Two varieties have been recognized, var. *diptera* with corollas 10–16 mm long, and var. *magniflora* Godfrey with corollas 18–32 mm long. Occasional. Floodplains, swamp margins, upland woods, ravines. To about 150 m elevation. Mar–May.

2.
Carolina Silverbell
Halesia tetraptera Ellis
[281]

Recognized by leaves oblong, ovate, to elliptic or oblong-ovate; by corollas shallowly 4-lobed, 14–25 mm long; by fruits 3–5 cm long with 4 prominent lengthwise wings. Shrubs or trees to 26 m tall by 1.3 m DBH. Bark on small branches with distinctive whitish streaks. Leaves 8–18 cm long, 4–10 cm wide, acute to acuminate, finely serrate. Styles about as long as corollas. Fruits oblong with 1–3 seeds. Sometimes planted as an ornamental; unripe fruits, which are sour, eaten by wildlife; limited use for lumber due to scarcity. Occasional overall, common in the sAppalachians. Rich soils, mostly along streams and in bottomlands; rich wooded slopes. To over 1700 m elevation in sAppalachians though rare above 1500 m. Mar–June.
H. carolina L. as used in most manuals. Use of the less familiar name, *H. tetraptera*, for this species instead of the commonly used *H. carolina* may cause confusion; but under the rules of nomenclature this change should be made since specimens of Silverbell from the locality where the original collection was made have corollas 14–18 mm long.

3.
Little Silverbell
Halesia carolina L.

Similar to the above species. It may be separated by the corollas being under 13 mm long, style markedly longer than corolla, the fruits under 3 cm long and tapered toward the base. Shrubs or trees to 9 m tall by 20 cm DBH. Rare. Floodplains and rich upland woods; uncommonly in dry sandy soils. To about 150 m elevation. Mar–Apr. Syn: *H. parviflora* Michx. as used in most manuals.

Index

EASTERN NORTH AMERICA

Areas in which this book is especially useful

centimeters (cm)